Introductory Discrete Mathematics

Introductory Discrete Mathematics

Edited by
Christopher Rhoades

CWILLFORD PRESS
www.willfordpress.com

Published by Willford Press,
118-35 Queens Blvd., Suite 400,
Forest Hills, NY 11375, USA

ISBN: 978-1-68285-492-1

Cataloging-in-Publication Data

Introductory discrete mathematics / edited by Christopher Rhoades.
     p. cm.
Includes bibliographical references and index.
ISBN 978-1-68285-492-1
1. Computer science--Mathematics. 2. Mathematics--Data processing. 3. Functional analysis.
I. Rhoades, Christopher.
QA76.9.M35 I58 2018
004.015 1--dc23

For information on all Willford Press publications
visit our website at www.willfordpress.com

WILLFORD PRESS

Contents

Preface

Discrete mathematics is defined as the study of mathematical structures which are not continuous. This includes integers, statements in logic, graphs, etc. The most important concepts studied under this subject are number theory, logic, probability, graph theory, algebra, set theory, information theory, combinatorics, geometry, game theory, topology, etc. This book provides comprehensive insights into the field of discrete mathematics. It aims to shed light on some of the unexplored aspects of the area. Different approaches, evaluations and methodologies have been included in the text. This book is an essential guide for both academicians and those who wish to pursue this discipline further.

Given below is the chapter wise description of the book:

Chapter 1- Discrete mathematics involves discrete elements with algebra and arithmetic. Logic, set theory, graph theory, probability, etc. are some of the main theories of discrete mathematics. This chapter is an overview of the subject matter incorporating all the major aspects of discrete mathematics.

Chapter 2- Functions are relations between sets of inputs and outputs where the input is somehow related to the output. The inputs are known as domain, the outputs codomain and the set that denotes their relationality as graph. This chapter elucidates the crucial theories and principles of discrete mathematics.

Chapter 3- Combinatorics deal with the study of finite or countable sets of discrete structures while generating functions deal with an infinite series of numbers. Pigeonhole principle, recurrence relation, formal power series, etc. are some of the main theories in this field. Discrete mathematics is best understood in confluence with the major topics listed in the following chapter.

Chapter 4- When two elements combine to form a third element, a group is formed. Groups are mostly used along with the use of polynomial equations. The aspects elucidated in this chapter are of vital importance, and provide a better understanding of discrete mathematics.

At the end, I would like to thank all those who dedicated their time and efforts for the successful completion of this book. I also wish to convey my gratitude towards my friends and family who supported me at every step.

Editor

Fundamentals of Discrete Mathematics

Discrete mathematics involves discrete elements with algebra and arithmetic. Logic, set theory, graph theory, probability, etc. are some of the main theories of discrete mathematics. This chapter is an overview of the subject matter incorporating all the major aspects of discrete mathematics.

Integer

An integer (from the Latin *integer* meaning "whole") is a number that can be written without a fractional component. For example, 21, 4, 0, and −2048 are integers, while 9.75, 5 ½, and $\sqrt{2}$ are not.

The set of integers consists of zero (0), the positive natural numbers (1, 2, 3, …), also called *whole numbers* or *counting numbers*, and their additive inverses (the negative integers, i.e., −1, −2, −3, …). This is often denoted by a boldface Z Z ("Z") or blackboard bold $\mathbb{Z}$ (Unicode U+2124 ℤ) standing for the German word *Zahlen*.

$\mathbb{Z}$ is a subset of the sets of rational numbers $\mathbb{Q}$, in turn a subset of the real numbers $\mathbb{R}$. Like the natural numbers, $\mathbb{Z}$ is countably infinite.

The integers form the smallest group and the smallest ring containing the natural numbers. In algebraic number theory, the integers are sometimes called rational integers to distinguish them from the more general algebraic integers. In fact, the (rational) integers are the algebraic integers that are also rational numbers.

Symbol

The symbol $\mathbb{Z}$ can be annotated to indicate various subsets, with varying usage amongst different authors: $\mathbb{Z}^+$ $\mathbb{Z}_+$ or $\mathbb{Z}^>$ for the positive integers, $\mathbb{Z}^{\geq}$ for non-negative integers, $\mathbb{Z}^{\neq}$ for non-zero integers, Some authors use $\mathbb{Z}^*$ for non-zero integers, others use it for non-negative integers. $\mathbb{Z}_n$ to mean the set of integers modulo n: {0, 1, 2, .. n-1}.

Algebraic Properties

Integers can be thought of as discrete, equally spaced points on an infinitely long number line. In the above, non-negative integers are shown in purple and negative integers in red.

Like the natural numbers, Z is closed under the operations of addition and multiplication, that is, the sum and product of any two integers is an integer. However, with the inclusion of the negative natural numbers, and, importantly, 0, Z (unlike the natural numbers) is also closed under subtraction. The integers form a unital ring which is the most basic one, in the following sense: for any unital ring, there is a unique ring homomorphism from the integers into this ring. This universal property, namely to be an initial object in the category of rings, characterizes the ring Z.

Z is not closed under division, since the quotient of two integers (e.g. 1 divided by 2), need not be an integer. Although the natural numbers are closed under exponentiation, the integers are not (since the result can be a fraction when the exponent is negative).

The following lists some of the basic properties of addition and multiplication for any integers a, b and c.

Properties of addition and multiplication on integers		
	Addition	**Multiplication**
Closure:	$a + b$ is an integer	$a \times b$ is an integer
Associativity:	$a + (b + c) = (a + b) + c$	$a \times (b \times c) = (a \times b) \times c$
Commutativity:	$a + b = b + a$	$a \times b = b \times a$
Existence of an identity element:	$a + 0 = a$	$a \times 1 = a$
Existence of inverse elements:	$a + (-a) = 0$	The only invertible rational integers (called units) are −1 and 1.
Distributivity:	$a \times (b + c) = (a \times b) + (a \times c)$ and $(a + b) \times c = (a \times c) + (b \times c)$	
No zero divisors:		If $a \times b = 0$, then $a = 0$ or $b = 0$ (or both)

In the language of abstract algebra, the first five properties listed above for addition say that Z under addition is an abelian group. As a group under addition, Z is a cyclic group, since every non-zero integer can be written as a finite sum $1 + 1 + ... + 1$ or $(-1) + (-1) + ... + (-1)$. In fact, Z under addition is the *only* infinite cyclic group, in the sense that any infinite cyclic group is isomorphic to Z.

The first four properties listed above for multiplication say that Z under multiplication is a commutative monoid. However, not every integer has a multiplicative inverse; e.g. there is no integer x such that $2x = 1$, because the left hand side is even, while the right hand side is odd. This means that Z under multiplication is not a group.

All the rules from the above property table, except for the last, taken together say that Z together with addition and multiplication is a commutative ring with unity. It is the prototype of all objects of such algebraic structure. Only those equalities of expressions are true in Z for all values of variables, which are true in any unital commutative ring. Note that certain non-zero integers map to zero in certain rings.

The lack of zero-divisors in the integers (last property in the table) means that the commutative ring Z is an integral domain.

The lack of multiplicative inverses, which is equivalent to the fact that Z is not closed under division, means that Z is *not* a field. The smallest field containing the integers as a subring is the field of rational numbers. The process of constructing the rationals from the integers can be mimicked to form the field of fractions of any integral domain. And back, starting from an algebraic number field (an extension of rational numbers), its ring of integers can be extracted, which includes Z as its subring.

Moreover, Z is a principal ideal.

Although ordinary division is not defined on Z, the division "with remainder" is defined on them. It is called Euclidean division and possesses the following important property: that is, given two integers a and b with $b \neq 0$, there exist unique integers q and r such that $a = q \times b + r$ and $0 \leq r < |b|$, where $|b|$ denotes the absolute value of b. The integer q is called the *quotient* and r is called the *remainder* of the division of a by b. The Euclidean algorithm for computing greatest common divisors works by a sequence of Euclidean divisions.

Again, in the language of abstract algebra, the above says that Z is a Euclidean domain. This implies that Z is a principal ideal domain and any positive integer can be written as the products of primes in an essentially unique way. This is the fundamental theorem of arithmetic.

Order-theoretic Properties

Z is a totally ordered set without upper or lower bound. The ordering of Z is given by: :... $-3 < -2 < -1 < 0 < 1 < 2 < 3 < ...$ An integer is *positive* if it is greater than zero and *negative* if it is less than zero. Zero is defined as neither negative nor positive.

The ordering of integers is compatible with the algebraic operations in the following way:

1.   if $a < b$ and $c < d$, then $a + c < b + d$.

2.   if $a < b$ and $0 < c$, then $ac < bc$.

It follows that Z together with the above ordering is an ordered ring.

The integers are the only nontrivial totally ordered abelian group whose positive elements are well-ordered. This is equivalent to the statement that any Noetherian valuation ring is either a field or a discrete valuation ring.

Construction

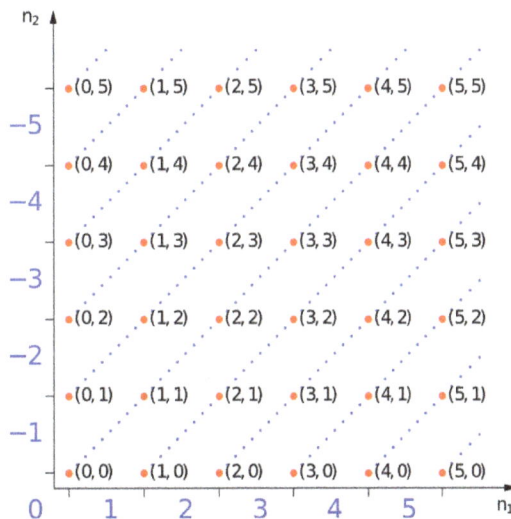

Red points represent ordered pairs of natural numbers. Linked red points are equivalence classes representing the blue integers at the end of the line.

In elementary school teaching, integers are often intuitively defined as the (positive) natural numbers, zero, and the negations of the natural numbers. However, this style of definition leads to many different cases (each arithmetic operation needs to be defined on each combination of types of integer) and makes it tedious to prove that these operations obey the laws of arithmetic. Therefore, in modern set-theoretic mathematics a more abstract construction, which allows one to define the arithmetical operations without any case distinction, is often used instead. The integers can thus be formally constructed as the equivalence classes of ordered pairs of natural numbers (a,b).

The intuition is that (a,b) stands for the result of subtracting b from a. To confirm our expectation that $1-2$ and $4-5$ denote the same number, we define an equivalence relation $\sim$ on these pairs with the following rule:

$$(a,b) \sim (c,d)$$

precisely when

$$a+d = b+c.$$

Addition and multiplication of integers can be defined in terms of the equivalent operations on the natural numbers; denoting by $[(a,b)]$ the equivalence class having (a,b) as a member, one has:

$$[(a,b)]+[(c,d)] := [(a+c,b+d)].$$

$$[(a,b)] \cdot [(c,d)] := [(ac+bd,ad+bc)].$$

The negation (or additive inverse) of an integer is obtained by reversing the order of the pair:

$$-[(a,b)] := [(b,a)].$$

Hence subtraction can be defined as the addition of the additive inverse:

$$[(a,b)]-[(c,d)] := [(a+d,b+c)].$$

The standard ordering on the integers is given by:

$$[(a,b)] < [(c,d)] \text{ iff } a+d < b+c.$$

It is easily verified that these definitions are independent of the choice of representatives of the equivalence classes.

Every equivalence class has a unique member that is of the form $(n,0)$ or $(0,n)$ (or both at once). The natural number n is identified with the class $[(n,0)]$ (in other words the natural numbers are embedded into the integers by map sending n to $[(n,0)]$), and the class $[(0,n)]$ is denoted $-n$ (this covers all remaining classes, and gives the class $[(0,0)]$ a second time since $-0 = 0$.

Thus, $[(a,b)]$ is denoted by

$$\begin{cases} a-b, & \text{if } a \geq b \\ -(b-a), & \text{if } a < b. \end{cases}$$

If the natural numbers are identified with the corresponding integers (using the embedding mentioned above), this convention creates no ambiguity.

This notation recovers the familiar representation of the integers as $\{..., -2, -1, 0, 1, 2, ...\}$.

Some examples are:

$$
\begin{aligned}
0 &= [(0,0)] & &= [(1,1)] = \cdots & &= [(k,k)] \\
1 &= [(1,0)] & &= [(2,1)] = \cdots & &= [(k+1,k)] \\
-1 &= [(0,1)] & &= [(1,2)] = \cdots & &= [(k,k+1)] \\
2 &= [(2,0)] & &= [(3,1)] = \cdots & &= [(k+2,k)] \\
-2 &= [(0,2)] & &= [(1,3)] = \cdots & &= [(k,k+2)].
\end{aligned}
$$

Computer Science

An integer is often a primitive data type in computer languages. However, integer data types can only represent a subset of all integers, since practical computers are of finite capacity. Also, in the common two's complement representation, the inherent definition of sign distinguishes between "negative" and "non-negative" rather than "negative, positive, and 0". (It is, however, certainly possible for a computer to determine whether an integer value is truly positive.) Fixed length integer approximation data types (or subsets) are denoted *int* or Integer in several programming languages (such as Algol68, C, Java, Delphi, etc.).

Variable-length representations of integers, such as bignums, can store any integer that fits in the computer's memory. Other integer data types are implemented with a fixed size, usually a number of bits which is a power of 2 (4, 8, 16, etc.) or a memorable number of decimal digits (e.g., 9 or 10).

Cardinality

The cardinality of the set of integers is equal to $\aleph_0$ (aleph-null). This is readily demonstrated by the construction of a bijection, that is, a function that is injective and surjective from Z to N. If N = {0, 1, 2, ...} then consider the function:

$$
f(x) = \begin{cases} 2|x|, & if\ x \leq 0 \\ 2x-1, & if\ x > 0. \end{cases}
$$

$\{... (-4,8)\ (-3,6)\ (-2,4)\ (-1,2)\ (0,0)\ (1,1)\ (2,3)\ (3,5)\ ...\}$

If N = {1, 2, 3, ...} then consider the function:

$$
g(x) = \begin{cases} 2|x|, & if\ x < 0 \\ 2x+1, & if\ x \geq 0. \end{cases}
$$

$\{... (-4,8)\ (-3,6)\ (-2,4)\ (-1,2)\ (0,1)\ (1,3)\ (2,5)\ (3,7)\ ...\}$

If the domain is restricted to Z then each and every member of Z has one and only one corresponding member of N and by the definition of cardinal equality the two sets have equal cardinality.

Set Theory

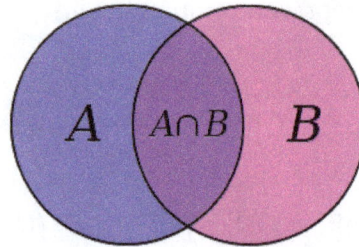

A Venn diagram illustrating the intersection of two sets.

Set theory is a branch of mathematical logic that studies sets, which informally are collections of objects. Although any type of object can be collected into a set, set theory is applied most often to objects that are relevant to mathematics. The language of set theory can be used in the definitions of nearly all mathematical objects.

The modern study of set theory was initiated by Georg Cantor and Richard Dedekind in the 1870s. After the discovery of paradoxes in naive set theory, such as the Russell's paradox, numerous axiom systems were proposed in the early twentieth century, of which the Zermelo–Fraenkel axioms, with the axiom of choice, are the best-known.

Set theory is commonly employed as a foundational system for mathematics, particularly in the form of Zermelo–Fraenkel set theory with the axiom of choice. Beyond its foundational role, set theory is a branch of mathematics in its own right, with an active research community. Contemporary research into set theory includes a diverse collection of topics, ranging from the structure of the real number line to the study of the consistency of large cardinals.

History

Georg Cantor.

Mathematical topics typically emerge and evolve through interactions among many researchers. Set theory, however, was founded by a single paper in 1874 by Georg Cantor: "On a Property of the Collection of All Real Algebraic Numbers".

Since the 5th century BC, beginning with Greek mathematician Zeno of Elea in the West and early Indian mathematicians in the East, mathematicians had struggled with the concept of infinity. Especially notable is the work of Bernard Bolzano in the first half of the 19th century. Modern understanding of infinity began in 1867–71, with Cantor's work on number theory. An 1872 meeting between Cantor and Richard Dedekind influenced Cantor's thinking and culminated in Cantor's 1874 paper.

Cantor's work initially polarized the mathematicians of his day. While Karl Weierstrass and Dedekind supported Cantor, Leopold Kronecker, now seen as a founder of mathematical constructivism, did not. Cantorian set theory eventually became widespread, due to the utility of Cantorian concepts, such as one-to-one correspondence among sets, his proof that there are more real numbers than integers, and the "infinity of infinities" ("Cantor's paradise") resulting from the power set operation. This utility of set theory led to the article "Mengenlehre" contributed in 1898 by Arthur Schoenflies to Klein's encyclopedia.

The next wave of excitement in set theory came around 1900, when it was discovered that some interpretations of Cantorian set theory gave rise to several contradictions, called antinomies or paradoxes. Bertrand Russell and Ernst Zermelo independently found the simplest and best known paradox, now called Russell's paradox: consider "the set of all sets that are not members of themselves", which leads to a contradiction since it must be a member of itself, and not a member of itself. In 1899 Cantor had himself posed the question "What is the cardinal number of the set of all sets?", and obtained a related paradox. Russell used his paradox as a theme in his 1903 review of continental mathematics in his *The Principles of Mathematics*.

In 1906 English readers gained the book *Theory of Sets of Points* by William Henry Young and his wife Grace Chisholm Young, published by Cambridge University Press.

The momentum of set theory was such that debate on the paradoxes did not lead to its abandonment. The work of Zermelo in 1908 and Abraham Fraenkel in 1922 resulted in the set of axioms ZFC, which became the most commonly used set of axioms for set theory. The work of analysts such as Henri Lebesgue demonstrated the great mathematical utility of set theory, which has since become woven into the fabric of modern mathematics. Set theory is commonly used as a foundational system, although in some areas category theory is thought to be a preferred foundation.

Basic Concepts and Notation

Set theory begins with a fundamental binary relation between an object o and a set A. If o is a member (or element) of A, the notation $o \in A$ is used. Since sets are objects, the membership relation can relate sets as well.

A derived binary relation between two sets is the subset relation, also called set inclusion. If all the members of set A are also members of set B, then A is a subset of B, denoted $A \subseteq B$. For example, $\{1, 2\}$ is a subset of $\{1, 2, 3\}$, and so is $\{2\}$ but $\{1, 4\}$ is not. As insinuated from this definition, a set is a subset of itself. For cases where this possibility is unsuitable or would make sense to be rejected, the term proper subset is defined. A is called a proper subset of B if and only if A is a subset of B, but A is not equal to B. Note also that 1 and 2 and 3 are members (elements) of set $\{1, 2, 3\}$, but are not subsets, and the subsets are, in turn, not as such members of the set.

Just as arithmetic features binary operations on numbers, set theory features binary operations on sets. The:

- Union of the sets A and B, denoted $A \cup B$, is the set of all objects that are a member of A, or B, or both. The union of $\{1, 2, 3\}$ and $\{2, 3, 4\}$ is the set $\{1, 2, 3, 4\}$.

- Intersection of the sets A and B, denoted $A \cap B$, is the set of all objects that are members of both A and B. The intersection of $\{1, 2, 3\}$ and $\{2, 3, 4\}$ is the set $\{2, 3\}$.

- Set difference of U and A, denoted $U \setminus A$, is the set of all members of U that are not members of A. The set difference $\{1, 2, 3\} \setminus \{2, 3, 4\}$ is $\{1\}$, while, conversely, the set difference $\{2, 3, 4\} \setminus \{1, 2, 3\}$ is $\{4\}$. When A is a subset of U, the set difference $U \setminus A$ is also called the complement of A in U. In this case, if the choice of U is clear from the context, the notation A^c is sometimes used instead of $U \setminus A$, particularly if U is a universal set as in the study of Venn diagrams.

- Symmetric difference of sets A and B, denoted $A \vartriangle B$ or $A \ominus B$, is the set of all objects that are a member of exactly one of A and B (elements which are in one of the sets, but not in both). For instance, for the sets $\{1, 2, 3\}$ and $\{2, 3, 4\}$, the symmetric difference set is $\{1, 4\}$. It is the set difference of the union and the intersection, $(A \cup B) \setminus (A \cap B)$ or $(A \setminus B) \cup (B \setminus A)$.

- Cartesian product of A and B, denoted $A \times B$, is the set whose members are all possible ordered pairs (a, b) where a is a member of A and b is a member of B. The cartesian product of $\{1, 2\}$ and $\{red, white\}$ is $\{(1, red), (1, white), (2, red), (2, white)\}$.

- Power set of a set A is the set whose members are all possible subsets of A. For example, the power set of $\{1, 2\}$ is $\{ \{\}, \{1\}, \{2\}, \{1, 2\} \}$.

Some basic sets of central importance are the empty set (the unique set containing no elements; occasionally called the *null set* though this name is ambiguous), the set of natural numbers, and the set of real numbers.

Some Ontology

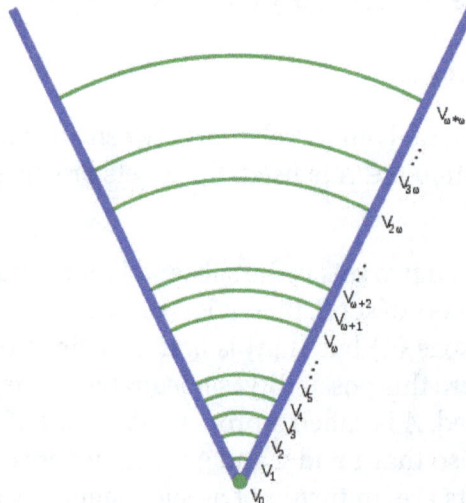

An initial segment of the von Neumann hierarchy.

A set is pure if all of its members are sets, all members of its members are sets, and so on. For example, the set {{}} containing only the empty set is a nonempty pure set. In modern set theory, it is common to restrict attention to the von Neumann universe of pure sets, and many systems of axiomatic set theory are designed to axiomatize the pure sets only. There are many technical advantages to this restriction, and little generality is lost, because essentially all mathematical concepts can be modeled by pure sets. Sets in the von Neumann universe are organized into a cumulative hierarchy, based on how deeply their members, members of members, etc. are nested. Each set in this hierarchy is assigned (by transfinite recursion) an ordinal number α, known as its rank. The rank of a pure set X is defined to be the least upper bound of all successors of ranks of members of X. For example, the empty set is assigned rank 0, while the set {{}} containing only the empty set is assigned rank 1. For each ordinal α, the set V_α is defined to consist of all pure sets with rank less than α. The entire von Neumann universe is denoted V.

Axiomatic Set Theory

Elementary set theory can be studied informally and intuitively, and so can be taught in primary schools using Venn diagrams. The intuitive approach tacitly assumes that a set may be formed from the class of all objects satisfying any particular defining condition. This assumption gives rise to paradoxes, the simplest and best known of which are Russell's paradox and the Burali-Forti paradox. Axiomatic set theory was originally devised to rid set theory of such paradoxes.

The most widely studied systems of axiomatic set theory imply that all sets form a cumulative hierarchy. Such systems come in two flavors, those whose ontology consists of:

- *Sets alone*. This includes the most common axiomatic set theory, Zermelo–Fraenkel set theory (ZFC), which includes the axiom of choice. Fragments of ZFC include:

    o Zermelo set theory, which replaces the axiom schema of replacement with that of separation;

    o General set theory, a small fragment of Zermelo set theory sufficient for the Peano axioms and finite sets;

    o Kripke–Platek set theory, which omits the axioms of infinity, powerset, and choice, and weakens the axiom schemata of separation and replacement.

- *Sets and proper classes*. These include Von Neumann–Bernays–Gödel set theory, which has the same strength as ZFC for theorems about sets alone, and Morse-Kelley set theory and Tarski–Grothendieck set theory, both of which are stronger than ZFC.

The above systems can be modified to allow urelements, objects that can be members of sets but that are not themselves sets and do not have any members.

The systems of New Foundations NFU (allowing urelements) and NF (lacking them) are not based on a cumulative hierarchy. NF and NFU include a "set of everything, " relative to which every set has a complement. In these systems urelements matter, because NF, but not NFU, produces sets for which the axiom of choice does not hold.

Systems of constructive set theory, such as CST, CZF, and IZF, embed their set axioms in intuitionistic instead of classical logic. Yet other systems accept classical logic but feature a nonstandard membership relation. These include rough set theory and fuzzy set theory, in which the value of an atomic formula embodying the membership relation is not simply True or False. The Boolean-valued models of ZFC are a related subject.

An enrichment of ZFC called Internal Set Theory was proposed by Edward Nelson in 1977.

Applications

Many mathematical concepts can be defined precisely using only set theoretic concepts. For example, mathematical structures as diverse as graphs, manifolds, rings, and vector spaces can all be defined as sets satisfying various (axiomatic) properties. Equivalence and order relations are ubiquitous in mathematics, and the theory of mathematical relations can be described in set theory.

Set theory is also a promising foundational system for much of mathematics. Since the publication of the first volume of *Principia Mathematica*, it has been claimed that most or even all mathematical theorems can be derived using an aptly designed set of axioms for set theory, augmented with many definitions, using first or second order logic. For example, properties of the natural and real numbers can be derived within set theory, as each number system can be identified with a set of equivalence classes under a suitable equivalence relation whose field is some infinite set.

Set theory as a foundation for mathematical analysis, topology, abstract algebra, and discrete mathematics is likewise uncontroversial; mathematicians accept that (in principle) theorems in these areas can be derived from the relevant definitions and the axioms of set theory. Few full derivations of complex mathematical theorems from set theory have been formally verified, however, because such formal derivations are often much longer than the natural language proofs mathematicians commonly present. One verification project, Metamath, includes human-written, computer-verified derivations of more than 12,000 theorems starting from ZFC set theory, first order logic and propositional logic.

Areas of Study

Set theory is a major area of research in mathematics, with many interrelated subfields.

Combinatorial Set Theory

Combinatorial set theory concerns extensions of finite combinatorics to infinite sets. This includes the study of cardinal arithmetic and the study of extensions of Ramsey's theorem such as the Erdős–Rado theorem.

Descriptive Set Theory

Descriptive set theory is the study of subsets of the real line and, more generally, subsets of Polish spaces. It begins with the study of pointclasses in the Borel hierarchy and extends to the study of more complex hierarchies such as the projective hierarchy and the Wadge hierarchy. Many properties of Borel sets can be established in ZFC, but proving these properties hold for more complicated sets requires additional axioms related to determinacy and large cardinals.

The field of effective descriptive set theory is between set theory and recursion theory. It includes the study of lightface pointclasses, and is closely related to hyperarithmetical theory. In many cases, results of classical descriptive set theory have effective versions; in some cases, new results are obtained by proving the effective version first and then extending ("relativizing") it to make it more broadly applicable.

A recent area of research concerns Borel equivalence relations and more complicated definable equivalence relations. This has important applications to the study of invariants in many fields of mathematics.

Fuzzy Set Theory

In set theory as Cantor defined and Zermelo and Fraenkel axiomatized, an object is either a member of a set or not. In fuzzy set theory this condition was relaxed by Lotfi A. Zadeh so an object has a *degree of membership* in a set, a number between 0 and 1. For example, the degree of membership of a person in the set of "tall people" is more flexible than a simple yes or no answer and can be a real number such as 0.75.

Inner Model Theory

An inner model of Zermelo–Fraenkel set theory (ZF) is a transitive class that includes all the ordinals and satisfies all the axioms of ZF. The canonical example is the constructible universe L developed by Gödel. One reason that the study of inner models is of interest is that it can be used to prove consistency results. For example, it can be shown that regardless of whether a model V of ZF satisfies the continuum hypothesis or the axiom of choice, the inner model L constructed inside the original model will satisfy both the generalized continuum hypothesis and the axiom of choice. Thus the assumption that ZF is consistent (has at least one model) implies that ZF together with these two principles is consistent.

The study of inner models is common in the study of determinacy and large cardinals, especially when considering axioms such as the axiom of determinacy that contradict the axiom of choice. Even if a fixed model of set theory satisfies the axiom of choice, it is possible for an inner model to fail to satisfy the axiom of choice. For example, the existence of sufficiently large cardinals implies that there is an inner model satisfying the axiom of determinacy (and thus not satisfying the axiom of choice).

Large Cardinals

A large cardinal is a cardinal number with an extra property. Many such properties are studied, including inaccessible cardinals, measurable cardinals, and many more. These properties typically imply the cardinal number must be very large, with the existence of a cardinal with the specified property unprovable in Zermelo-Fraenkel set theory.

Determinacy

Determinacy refers to the fact that, under appropriate assumptions, certain two-player games of perfect information are determined from the start in the sense that one player must have a winning strategy. The existence of these strategies has important consequences in descriptive set theory, as the assumption that a broader class of games is determined often implies that a broader class of sets will have a topological property. The axiom of determinacy (AD) is an important object of study; although incompatible with the axiom of choice, AD implies that all subsets of the real line are well behaved (in particular, measurable and with the perfect set property). AD can be used to prove that the Wadge degrees have an elegant structure.

Forcing

Paul Cohen invented the method of forcing while searching for a model of ZFC in which the continuum hypothesis fails, or a model of ZF in which the axiom of choice fails. Forcing adjoins to some given model of set theory additional sets in order to create a larger model with properties determined (i.e. "forced") by the construction and the original model. For example, Cohen's construction adjoins additional subsets of the natural numbers without changing any of the cardinal numbers of the original model. Forcing is also one of two methods for proving relative consistency by finitistic methods, the other method being Boolean-valued models.

Cardinal Invariants

A cardinal invariant is a property of the real line measured by a cardinal number. For example, a well-studied invariant is the smallest cardinality of a collection of meagre sets of reals whose union is the entire real line. These are invariants in the sense that any two isomorphic models of set theory must give the same cardinal for each invariant. Many cardinal invariants have been studied, and the relationships between them are often complex and related to axioms of set theory.

Set-theoretic Topology

Set-theoretic topology studies questions of general topology that are set-theoretic in nature or that require advanced methods of set theory for their solution. Many of these theorems are independent of ZFC, requiring stronger axioms for their proof. A famous problem is the normal Moore space question, a question in general topology that was the subject of intense research. The answer to the normal Moore space question was eventually proved to be independent of ZFC.

Objections to Set Theory as a Foundation for Mathematics

From set theory's inception, some mathematicians have objected to it as a foundation for mathematics. The most common objection to set theory, one Kronecker voiced in set theory's earliest years, starts from the constructivist view that mathematics is loosely related to computation. If this view is granted, then the treatment of infinite sets, both in naive and in axiomatic set theory, introduces into mathematics methods and objects that are not computable even in principle. The feasibility of constructivism as a substitute foundation for mathematics was greatly increased by Errett Bishop's influential book *Foundations of Constructive Analysis*.

A different objection put forth by Henri Poincaré is that defining sets using the axiom schemas of specification and replacement, as well as the axiom of power set, introduces impredicativity, a type of circularity, into the definitions of mathematical objects. The scope of predicatively founded mathematics, while less than that of the commonly accepted Zermelo-Fraenkel theory, is much greater than that of constructive mathematics, to the point that Solomon Feferman has said that "all of scientifically applicable analysis can be developed [using predicative methods]".

Ludwig Wittgenstein condemned set theory. He wrote that "set theory is wrong", since it builds on the "nonsense" of fictitious symbolism, has "pernicious idioms", and that it is nonsensical to talk about "all numbers". Wittgenstein's views about the foundations of mathematics were later criticised by Georg Kreisel and Paul Bernays, and investigated by Crispin Wright, among others.

Category theorists have proposed topos theory as an alternative to traditional axiomatic set theory. Topos theory can interpret various alternatives to that theory, such as constructivism, finite set theory, and computable set theory. Topoi also give a natural setting for forcing and discussions of the independence of choice from ZF, as well as providing the framework for pointless topology and Stone spaces.

An active area of research is the univalent foundations and related to it homotopy type theory. Here, sets may be defined as certain kinds of types, with universal properties of sets arising from higher inductive types. Principles such as the axiom of choice and the law of the excluded middle appear in a spectrum of different forms, some of which can be proven, others which correspond to the classical notions; this allows for a detailed discussion of the effect of these axioms on mathematics.

Basic Set Theory

We have already seen examples of sets, such as $\mathbb{N}, \mathbb{Z}, \mathbb{Q}, \mathbb{R}$ and $\mathbb{C}$. For example, one can also look at the following sets.

Example:

1. $\{1, 3, 5, 7, \ldots\}$, the set of odd natural numbers.

2. $\{0, 2, 4, 6, \ldots\}$, the set of even natural numbers.

3. $\{\ldots, -5, -3, -1, 1, 3, 5, \ldots\}$, the set of odd integers.

4. $\{\ldots, -6, -4, -2, 0, 2, 4, 6, \ldots\}$, the set of even integers.

5. $\{0, 1, 2, \ldots, 10\}$.

6. $\{1, 2, \ldots, 10\}$.

7. $\mathbb{Q}^+ = \{x \in \mathbb{Q} : x > 0\}$, the set of positive rational numbers.

8. $\mathbb{R}^+ = \{x \in \mathbb{R} : x > 0\}$, the set of positive real numbers.

9. $\mathbb{Q}^* = \{x \in \mathbb{Q} : x \neq 0\}$, the set of non-zero rational numbers.

10. $\mathbb{R}^* = \{x \in \mathbb{R} : x \neq 0\}$, the set of non-zero real numbers.

We observe that the sets that appear have been obtained by picking certain elements from the sets $\mathbb{N}, \mathbb{Z}, \mathbb{Q}, \mathbb{R}$ or $\mathbb{C}$. These sets are example of what are called "subsets of a set", which we define next. We also define certain operations on sets.

Definition (Subset, Complement, Union, Intersection): 1. Let A be a set. If B is a set such that each element of B is also an element of the set A, then B is said to be a subset of the set A, denoted $B \subseteq A$.

2. Two sets A and B are said to be equal if $A \subseteq B$ and $B \subseteq A$, denoted A = B.

3. Let A be a subset of a set Ω. Then the complement of A in Ω, denoted A', is a set that contains every element of Ω that is not an element of A. Specifically, $A' = \{x \in \Omega : x \notin A\}$.

4. Let A and B be two subsets of a set Ω. Then their

(a) union, denoted $A \cup B$, is the set that exactly contains all the elements of A and all the elements of B. To be more precise, $A \cup B = \{x \in \Omega : x\, A\, or\, x \in B\}$.

(b) intersection, denoted $A \cap B$, is the set that exactly contains those elements of A that are also elements of B. To be more precise, $A \cap B = \{x \in \Omega : x\, A\, and\, x \in B\}$.

Example

1. Let A be a set. Then $A \subseteq A$.

2. The empty set is a subset of every set.

3. Observe that $\mathbb{N} \subseteq \mathbb{Z} \subseteq \mathbb{Q} \subseteq \mathbb{R} \subseteq \mathbb{C}$.

4. As mentioned earlier, all examples that appear in Example 1.1.1 are subsets of one or more sets from $\mathbb{N}, \mathbb{Z}, \mathbb{Q}, \mathbb{R}$ and $\mathbb{C}$.

5. Let A be the set of odd integers and B be the set of even integers. Then $A \cap B = \varnothing$ and $A \cup B = \mathbb{Z}$. Thus, it also follows that the complement of A, in Z, equals B and vice-versa.

6. Let $A = \{\{b,c\}, \{\{b\}, \{c\}\}\}$ and $B = \{a,b,c\}$ be subsets of a set Ω. Then $A \cap B = \varnothing$ and $A \cup B = \{a,b,c,\{b,c\}, \{\{b\}, \{c\}\}\}$.

Definition (Cardinality): A set A is said to have finite cardinality, denoted $|A|$, if the number of distinct elements in A is finite, else the set A is said to have infinite cardinality.

Example

1. The cardinality of the empty set equals 0. That is, $|\varnothing| = 0$.

2. Fix a positive integer n and consider the set $A = \{1, 2, ..., n\}$. Then $|A| = n$.

3. Let $S = \{2x \in \mathbb{Z} : x \in \mathbb{Z}\}$. Then S is the set of even integers and it's cardinality is infinite.

4. Let $A = \{a_1, a_2, ..., a_m\}$ and $B = \{b_1, b_2, ..., b_n\}$ be two finite subsets of α set Ω, with $|A| = m$ and $|B| = n$. Also, assume that $A \cap B = \varnothing$. Then, by definition it follows that $A \cup B = \{a_1, b_2, ...a_m, b_1, b_2, ..., b_n\}$ and hence $|A \cup B| = |A| + |B|$.

5. Let $A = \{a_1, a_2, ..., a_m\}$ and $B = \{b_1, b_2, ..., b_n\}$ be two finite subsets of α set Ω. Then $|A \cup B| = |A| + |B| - |A \cap B|$. Observe that Example 1.1.5.4 is a particular case of this result, when $A \cap B = \varnothing$.

6. Let $A = \{\{a_1\}, \{a_2\}, ..., \{a_m\}\}$ be a collection of singletons of a set Ω. Now choose an element $\alpha \in \Omega$ such that $a \neq a_i$ for any $i, 1 \leq i \leq n$. Then verify that the set $B = \{S \cup \{a\} : S \in A\}$ equals $\{\{a, a_1\}, \{a, a_2\}, ..., \{a., a_m\}\}$. Also, observe that $A \cap B = \varnothing$ and $|B| = |A|$.

Definition (Power Set): Let A be a subset of α set Ω. Then the set that contains all subsets of A is called the power set of A and is denoted by $P(A)$ or 2^A.

Example.

1. Let $A = \varnothing$. Then $P(\varnothing) = \{\varnothing, A\} = \{\varnothing\}$.

2. Let A = $\{\varnothing\}$. Then P(A) = $\{\varnothing, A\}$ = $\{\varnothing, \{\varnothing\}\}$.

3. Let A = {a, b, c}. Then P(A) = {∅, {a}, {b}, {c}, {a, b}, {a, c}, {b, c}, {a, b, c}}.

4. Let A = {{b, c}, {{b}, {c}}}. Then P(A) = {∅, {{b, c}}, {{{b}, {c}}}, {{b, c}, {{b}, {c}}} }.

Mathematical Induction

Mathematical induction can be informally illustrated by reference to
the sequential effect of falling dominoes.

Mathematical induction is a mathematical proof technique used to prove a given statement about any well-ordered set. Most commonly, it is used to establish statements for the set of all natural numbers.

Mathematical induction is a form of direct proof, usually done in two steps. When trying to prove a given statement for a set of natural numbers, the first step, known as the base case, is to prove the given statement for the first natural number. The second step, known as the inductive step, is to prove that, if the statement is assumed to be true for any one natural number, then it must be true for the next natural number as well. Having proved these two steps, the rule of inference establishes the statement to be true for all natural numbers. In common terminology, using the stated approach is referred to as using the *Principle of mathematical induction*.

The method can be extended to prove statements about more general well-founded structures, such as trees; this generalization, known as structural induction, is used in mathematical logic and computer science. Mathematical induction in this extended sense is closely related to recursion. Mathematical induction, in some form, is the foundation of all correctness proofs for computer programs.

Although its name may suggest otherwise, mathematical induction should not be misconstrued as a form of inductive reasoning. Mathematical induction is an inference rule used in proofs. In mathematics, proofs including those using mathematical induction are examples of deductive reasoning, and inductive reasoning is excluded from proofs.

History

In 370 BC, Plato's Parmenides may have contained an early example of an implicit inductive proof. The earliest implicit traces of mathematical induction may be found in Euclid's proof that the number of primes is infinite and in Bhaskara's "cyclic method". An opposite iterated technique, counting *down* rather than up, is found in the Sorites paradox, where it was argued that if 1,000,000 grains of sand formed a heap, and removing one grain from a heap left it a heap, then a single grain of sand (or even no grains) forms a heap.

An implicit proof by mathematical induction for arithmetic sequences was introduced in the *al-Fakhri* written by al-Karaji around 1000 AD, who used it to prove the binomial theorem and properties of Pascal's triangle.

None of these ancient mathematicians, however, explicitly stated the inductive hypothesis. Another similar case (contrary to what Vacca has written, as Freudenthal carefully showed) was that of Francesco Maurolico in his *Arithmeticorum libri duo* (1575), who used the technique to prove that the sum of the first n odd integers is n^2. The first explicit formulation of the principle of induction was given by Pascal in his *Traité du triangle arithmétique* (1665). Another Frenchman, Fermat, made ample use of a related principle, indirect proof by infinite descent. The inductive hypothesis was also employed by the Swiss Jakob Bernoulli, and from then on it became more or less well known. The modern rigorous and systematic treatment of the principle came only in the 19th century, with George Boole, Augustus de Morgan, Charles Sanders Peirce, Giuseppe Peano, and Richard Dedekind.

Description

The simplest and most common form of mathematical induction infers that a statement involving a natural number n holds for all values of n. The proof consists of two steps:

1. The basis (base case): prove that the statement holds for the first natural number n. Usually, $n = 0$ or $n = 1$, rarely, $n = -1$ (although not a natural number, the extension of the natural numbers to −1 is still a well-ordered set).

2. The inductive step: prove that, if the statement holds for some natural number n, then the statement holds for $n + 1$.

The hypothesis in the inductive step that the statement holds for some n is called the induction hypothesis (or inductive hypothesis). To perform the inductive step, one assumes the induction hypothesis and then uses this assumption to prove the statement for $n + 1$.

Whether $n = 0$ or $n = 1$ depends on the definition of the natural numbers. If 0 is considered a natural number, as is common in the fields of combinatorics and mathematical logic, the base case is given by $n = 0$. If, on the other hand, 1 is taken as the first natural number, then the base case is given by $n = 1$.

Examples

Mathematical induction can be used to prove that the following statement, $P(n)$, holds for all natural numbers n.

$$0 + 1 + 2 + \cdots + n = \frac{n(n+1)}{2}.$$

$P(n)$ gives a formula for the sum of the natural numbers less than or equal to number n. The proof that $P(n)$ is true for each natural number n proceeds as follows.

Basis: Show that the statement holds for $n = 0$.

$P(0)$ amounts to the statement:

$$0 = \frac{0 \cdot (0+1)}{2}.$$

In the equation, the only term is 0, and so the left-hand side is simply equal to 0.

In the right-hand side of the equation, $0 \cdot (0 + 1)/2 = 0$.

The two sides are equal, so the statement is true for $n = 0$. Thus it has been shown that $P(0)$ holds.

Inductive step: Show that *if* $P(k)$ holds, then also $P(k + 1)$ holds. This can be done as follows.

Assume $P(k)$ holds (for some unspecified value of k). It must then be shown that $P(k + 1)$ holds, that is:

$$(0 + 1 + 2 + \cdots + k) + (k + 1) = \frac{(k+1)((k+1)+1)}{2}.$$

Using the induction hypothesis that $P(k)$ holds, the left-hand side can be rewritten to:

$$\frac{k(k+1)}{2} + (k+1).$$

Algebraically:

$$\frac{k(k+1)}{2} + (k+1) = \frac{k(k+1) + 2(k+1)}{2}$$
$$= \frac{(k+1)(k+2)}{2}$$
$$= \frac{(k+1)((k+1)+1)}{2}$$

thereby showing that indeed $P(k + 1)$ holds.

Since both the basis and the inductive step have been performed, by mathematical induction, the statement $P(n)$ holds for all natural numbers n. Q.E.D.

Axiom of Induction

Mathematical induction as an inference rule can be formalized as a second-order axiom. The *axiom of induction* is, in logical symbols,

$$\forall P.[[P(0) \wedge \forall (k \in \mathbb{N}).[P(k) \Rightarrow P(k+1)]] \Rightarrow \forall (n \in \mathbb{N}).P(n)]$$

where P is any predicate and k and n are both natural numbers.

In words, the basis $P(0)$ and the inductive step (namely, that the inductive hypothesis $P(k)$ implies $P(k + 1)$) together imply that $P(n)$ for any natural number n. The axiom of induction asserts that the validity of inferring that $P(n)$ holds for any natural number n from the basis and the inductive step.

Note that the first quantifier in the axiom ranges over *predicates* rather than over individual numbers. This is a second-order quantifier, which means that this axiom is stated in second-order logic. Axiomatizing arithmetic induction in first-order logic requires an axiom schema containing a separate axiom for each possible predicate.

Variants

In practice, proofs by induction are often structured differently, depending on the exact nature of the property to be proved.n

Induction Basis Other than 0 or 1

If one wishes to prove a statement not for all natural numbers but only for all numbers greater than or equal to a certain number b, then the proof by induction consists of:

1. Showing that the statement holds when $n = b$.

2. Showing that if the statement holds for $n = m \geq b$ then the same statement also holds for $n = m + 1$.

This can be used, for example, to show that $n^2 \geq 3n$ for $n \geq 3$. A more substantial example is a proof that

$$\frac{n^n}{3^n} < n! < \frac{n^n}{2^n} \text{ for } n \geq 6.$$

In this way, one can prove that $P(n)$ holds for all $n \geq 1$, or even $n \geq -5$. This form of mathematical induction is actually a special case of the previous form because if the statement to be proved is $P(n)$ then proving it with these two rules is equivalent with proving $P(n + b)$ for all natural numbers n with the first two steps.

Induction Basis Equal to 2

In mathematics, many standard functions, including operations such as "+" and relations such as "=", are binary, meaning that they take two arguments. Often these functions possess properties that implicitly extend them to more than two arguments. For example, once addition $a + b$ is defined and is known to satisfy the associativity property $(a + b) + c = a + (b + c)$, then the ternary addition $a + b + c$ makes sense, either as $(a + b) + c$ or as $a + (b + c)$. Similarly, many axioms and theorems in mathematics are stated only for the binary versions of mathematical operations and relations, and implicitly extend to higher-arity versions.

Suppose that one wishes to prove a statement about an n-ary operation implicitly defined from a binary operation, using mathematical induction on n. In this case it is natural to take 2 for the induction basis.

Example: Product Rule for the Derivative

In this example, the binary operation in question is multiplication (of functions). The usual product rule for the derivative taught in calculus states:

$$(fg)' = f'g + g'f.$$

or in logarithmic derivative form

$$(fg)' / (fg) = f' / f + g' / g.$$

This can be generalized to a product of n functions. One has

$$(f_1 f_2 f_3 \cdots f_n)'$$
$$= (f_1' f_2 f_3 \cdots f_n) + (f_1 f_2' f_3 \cdots f_n) + (f_1 f_2 f_3' \cdots f_n) + \cdots + (f_1 f_2 \cdots f_{n-1} f_n').$$

or in logarithmic derivative form

$$(f_1 f_2 f_3 \cdots f_n)' / (f_1 f_2 f_3 \cdots f_n)$$
$$= (f_1' / f_1) + (f_2' / f_2) + (f_3' / f_3) + \cdots + (f_n' / f_n).$$

In each of the n terms of the usual form, just one of the factors is a derivative; the others are not.

When this general fact is proved by mathematical induction, the $n = 0$ case is trivial, $(1)' = 0$ (since the empty product is 1, and the empty sum is 0). The $n = 1$ case is also trivial, $f_1' = f_1'$. And for each $n \geq 3$, the case is easy to prove from the preceding $n - 1$ case. The real difficulty lies in the $n = 2$ case, which is why that is the one stated in the standard product rule.

Example: Forming Dollar Sums by Coins

Induction can be used to prove that any dollar sum greater than 12 can be formed by the combination of 4 and 5 dollar coins. In more precise terms, for any the total dollar sum $n \geq 12$ there exist natural numbers a, b such that $n = 4a + 5b$. The statement to be shown is thus:

$$S(n): n \geq 12 \Rightarrow \exists a, b \in \mathbb{N}. n = 4a + 5b$$

Basis: Showing that $S(k)$ holds for $k = 12$ is trivial: let $a = 3$ and $b = 0$. Then, $4 \cdot 3 + 5 \cdot 0 = 12$.

Inductive Step: Given that $S(k)$ holds for some value of k (*inductive hypothesis*), prove that $S(k+1)$ holds, too. That is, given that $k = 4a + 5b$ for some natural numbers a, b, prove that there exist natural numbers a_1, b_1 such that $k + 1 = 4a_1 + 5b_1$.

By some algebraic manipulation and by assumption, we see that

$$
\begin{aligned}
k &= 4a + 5b \\
k + 1 &= 4a + 5b + 1 \\
&= 4a + 5b - 4 + 5 \\
&= 4(a - 1) + 5(b + 1) \\
&= 4a_1 + 5b_1
\end{aligned}
$$

where a_1, b_1 are natural numbers, provided that $a \geq 1$.

This shows that to add 1 to the total sum—any sum whatsoever, so long as it is greater than 12—it is sufficient to remove a single 4 dollar coin and add a 5 dollar coin. However, the proof above would fail if we had no 4 dollar coin.

So it remains to prove the case $a = 0$.

$$
\begin{aligned}
k &= 5b \\
k + 1 &= 5b + 1 \\
&= 5b + 1 + 15 - 15 \\
&= 5(b - 3) + 1 + 15 \\
&= 5(b - 3) + 16 \\
&= 5(b - 3) + 4 \cdot 4 \\
&= 5(b - 3) + 4a_2 \\
&= 4a_2 + 5b_2
\end{aligned}
$$

where a_2, b_2 are natural numbers, provided that $b \geq 3$.

The above shows that if we had no 4 dollar coins, we would add 1 to the sum by taking away three 5 dollar coins and adding four 4 dollar coins. By initial statement, the total sum must be no less than 12, implying a minimum of three 5 dollar coins.

Thus, with the inductive step, we have shown that $S(k)$ implies $S(k+1)$ for all natural numbers $k \geq 12$, and the proof is complete. Q.E.D.

Induction on More than one Counter

It is sometimes desirable to prove a statement involving two natural numbers, n and m, by iterating the induction process. That is, one performs a basis step and an inductive step for n, and in each of those performs a basis step and an inductive step for m. More complicated arguments involving three or more counters are also possible.

Infinite Descent

The method of infinite descent was one of Pierre de Fermat's favorites. This method of proof can assume several slightly different forms. For example, it might begin by showing that if a statement is true for a natural number n it must also be true for some smaller natural number m ($m < n$). Using mathematical induction (implicitly) with the inductive hypothesis being that the statement is false for all natural numbers less than or equal to m, one may conclude that the statement cannot be true for any natural number n.

Although this particular form of infinite-descent proof is clearly a mathematical induction, whether one holds all proofs "by infinite descent" to be mathematical inductions depends on how one defines the term "proof by infinite descent." One might, for example, use the term to apply to proofs in which the well-ordering of the natural numbers is assumed, but not the principle of induction. Such, for example, is the usual proof that 2 has no rational square root.

Prefix induction

The most common form of induction requires proving that

$$\forall k(P(k) \to P(k+1))$$

or equivalently

$$\forall k(P(k-1) \to P(k))$$

whereupon the induction principle "automates" n applications of this inference in getting from $P(0)$ to $P(n)$. This could be called "predecessor induction" because each step proves something about a number from something about that number's predecessor.

A variant of interest in computational complexity is "prefix induction", in which one needs to prove

$$\forall k(P(k) \to P(2k) \wedge P(2k+1))$$

or equivalently

$$\forall k\left(P\left(\left\lfloor \frac{k}{2} \right\rfloor \right) \to P(k) \right)$$

The induction principle then "automates" $\log n$ applications of this inference in getting from $P(0)$ to $P(n)$. (It is called "prefix induction" because each step proves something about a number from something about the "prefix" of that number formed by truncating the low bit of its binary representation.)

If traditional predecessor induction is interpreted computationally as an n-step loop, prefix induction corresponds to a $\log n$-step loop, and thus proofs using prefix induction are "more feasibly constructive" than proofs using predecessor induction.

Predecessor induction can trivially simulate prefix induction on the same statement. Prefix induction can simulate predecessor induction, but only at the cost of making the statement more syntactically complex (adding a bounded universal quantifier), so the interesting results relating prefix induction to polynomial-time computation depend on excluding unbounded quantifiers entirely, and limiting the alternation of bounded universal and existential quantifiers allowed in the statement.

One could take it a step farther to "prefix of prefix induction": one must prove

$$\forall k \left(P\left(\left\lfloor \sqrt{k} \right\rfloor\right) \rightarrow P(k) \right)$$

whereupon the induction principle "automates" $\log \log n$ applications of this inference in getting from $P(0)$ to $P(n)$. This form of induction has been used, analogously, to study log-time parallel computation.

Complete Induction

Another variant, called complete induction, course of values induction or strong induction (in contrast to which the basic form of induction is sometimes known as weak induction) makes the inductive step easier to prove by using a stronger hypothesis: one proves the statement P(m + 1) under the assumption that P(n) holds for all natural n less than m + 1; by contrast, the basic form only assumes P(m). The name "strong induction" does not mean that this method can prove more than "weak induction", but merely refers to the stronger hypothesis used in the inductive step; in fact the two methods are equivalent, as explained below. In this form of complete induction one still has to prove the base case, P(0), and it may even be necessary to prove extra base cases such as P(1) before the general argument applies, as in the example below of the Fibonacci number F_n.

Although the form just described requires one to prove the base case, this is unnecessary if one can prove P(m) (assuming P(n) for all lower n) for all $m \geq 0$. This is a special case of transfinite induction as described below. In this form the base case is subsumed by the case $m = 0$, where P(0) is proved with no other P(n) assumed; this case may need to be handled separately, but sometimes the same argument applies for $m = 0$ and $m > 0$, making the proof simpler and more elegant. In this method it is, however, vital to ensure that the proof of P(m) does not implicitly assume that $m > 0$, e.g. by saying "choose an arbitrary $n < m$" or assuming that a set of m elements has an element.

Complete induction is equivalent to ordinary mathematical induction as described above, in the sense that a proof by one method can be transformed into a proof by the other. Suppose there is a proof of P(n) by complete induction. Let Q(n) mean "P(m) holds for all m such that $0 \leq m \leq n$". Then Q(n) holds for all n if and only if P(n) holds for all n, and our proof of P(n) is easily transformed into a proof of Q(n) by (ordinary) induction. If, on the other hand, P(n) had been proven by ordinary induction, the proof would already effectively be one by complete induction: P(0) is proved in the base case, using no assumptions, and P(n + 1) is proved in the inductive step, in which one may assume all earlier cases but need only use the case P(n).

Example: Fibonacci Numbers

Complete induction is most useful when several instances of the inductive hypothesis are required for each inductive step. For example, complete induction can be used to show that

$$F_n = \frac{\varphi^n - \psi^n}{\varphi - \psi}$$

where F_n is the nth Fibonacci number, $\varphi = (1 + \sqrt{5})/2$ (the golden ratio) and $\psi = (1 - \sqrt{5})/2$ are the roots of the polynomial $x^2 - x - 1$. By using the fact that $F_{n+2} = F_{n+1} + F_n$ for each $n \in \mathbb{N}$, the identity above can be verified by direct calculation for F_{n+2} if one assumes that it already holds for both F_{n+1} and F_n. To complete the proof, the identity must be verified in the two base cases $n = 0$ and $n = 1$.

Example: Prime Factorization

Another proof by complete induction uses the hypothesis that the statement holds for *all* smaller n more thoroughly. Consider the statement that "every natural number greater than 1 is a product of (one or more) prime numbers", and assume that for a given $m > 1$ it holds for all smaller $n > 1$. If m is prime then it is certainly a product of primes, and if not, then by definition it is a product: $m = n_1 n_2$, where neither of the factors is equal to 1; hence neither is equal to m, and so both are smaller than m. The induction hypothesis now applies to n_1 and n_2, so each one is a product of primes. Thus m is a product of products of primes; i.e. a product of primes.

Example: Dollar Sums Revisited

We shall look to prove the same example as above, this time with a variant called *strong induction*. The statement remains the same:

$$S(n): n \geq 12 \Rightarrow \exists a, b \in \mathbb{N}. n = 4a + 5b$$

However, there will be slight differences with the structure and assumptions of the proof. Let us begin with the basis.

Basis: Show that $S(k)$ holds for $k = 12, 13, 14, 15$.

$$4 \cdot 3 + 5 \cdot 0 = 12$$
$$4 \cdot 2 + 5 \cdot 1 = 13$$
$$4 \cdot 1 + 5 \cdot 2 = 14$$
$$4 \cdot 0 + 5 \cdot 3 = 15$$

The basis holds.

Inductive Hypothesis: Given some $j > 15$ such that $S(m)$ holds for all m with $12 \leq m < j$.

Inductive Step: Prove that $S(j)$ holds.

Choosing $m = j - 4$, and observing that $15 < j \Rightarrow 12 \leq j - 4 < j$ shows that $S(j-4)$ holds, by inductive hypothesis. That is, the sum $j - 4$ can be formed by some combination of 4 and 5 dollar coins. Then, simply adding a 4 dollar coin to that combination yields the sum j. That is, $S(j)$ holds. Q.E.D.

Transfinite Induction

The last two steps can be reformulated as one step:

- Showing that if the statement holds for all $n < m$ then the same statement also holds for $n = m$.

This form of mathematical induction is not only valid for statements about natural numbers, but for statements about elements of any well-founded set, that is, a set with an irreflexive relation $<$ that contains no infinite descending chains.

This form of induction, when applied to ordinals (which form a well-ordered and hence well-founded class), is called *transfinite induction*. It is an important proof technique in set theory, topology and other fields.

Proofs by transfinite induction typically distinguish three cases:

1. when m is a minimal element, i.e. there is no element smaller than m

2. when m has a direct predecessor, i.e. the set of elements which are smaller than m has a largest element

3. when m has no direct predecessor, i.e. m is a so-called limit-ordinal

Strictly speaking, it is not necessary in transfinite induction to prove the basis, because it is a vacuous special case of the proposition that if P is true of all $n < m$, then P is true of m. It is vacuously true precisely because there are no values of $n < m$ that could serve as counterexamples.

Equivalence with the Well-ordering Principle

The principle of mathematical induction is usually stated as an axiom of the natural numbers. However, it can be proved from the well-ordering principle. Indeed, suppose the following:

- The set of natural numbers is well-ordered.

- Every natural number is either zero, or $n + 1$ for some natural number n.

- For any natural number n, $n + 1$ is greater than n.

To derive simple induction from these axioms, one must show that if P(n) is some proposition predicated of n, and if:

- P(0) holds and

- whenever P(k) is true then P($k + 1$) is also true

then P(n) holds for all n.

Proof. Let S be the set of all natural numbers for which P(n) is false. Let us see what happens if one asserts that S is nonempty. Well-ordering tells us that S has a least element, say t. Moreover, since P(0) is true, t is not 0. Since every natural number is either zero or some $n + 1$, there is some natural number n such that $n + 1 = t$. Now n is less than t, and t is the least element of S. It follows that n is not in S, and so P(n) is true. This means that P($n + 1$) is true, and so P(t) is true. This is a contradiction, since t was in S. Therefore, S is empty.It can also be proved that induction, given the other axioms, implies the well-ordering principle.

Proof. Suppose there exists a non-empty set, S, of naturals with no least element. Let $P(k)$ be the assertion that k is not in S. Then $P(0)$ is true for if it were false then 0 is the least element of S. Furthermore, suppose $P(1)$, $P(2)$,..., $P(k)$ is true. Then if $P(k+1)$ is false $k+1$ is in S, thus it is the minimal element is S, a contradiction. Thus $P(k+1)$ is true. Therefore, by the induction axiom S is empty, a contradiction.

Example of Error in the Inductive Step

This example demonstrated a subtle error in the proof of the inductive step.

Joel E. Cohen proposed the following argument, which purports to prove by mathematical induction that all horses are of the same color:

- Basis: In a set of only *one* horse, there is only one color.

- Induction step: Assume as induction hypothesis that within any set of n horses, there is only one color. Now look at any set of $n + 1$ horses. Number them: 1, 2, 3, ..., n, $n + 1$. Consider the sets $\{1, 2, 3, ..., n\}$ and $\{2, 3, 4, ..., n + 1\}$. Each is a set of only n horses, therefore within each there is only one color. But the two sets overlap, so there must be only one color among all $n + 1$ horses.

The basis case $n = 1$ is trivial (as any horse is the same color as itself), and the inductive step is correct in all cases $n > 1$. However, the logic of the inductive step is incorrect for $n = 1$, because the statement that «the two sets overlap» is false (there are only $n + 1 = 2$ horses prior to either removal, and after removal the sets of one horse each do not overlap).

Well Ordering Principle and the Principle of Mathematical Induction

Axiom (Well-Ordering Principle). Every non-empty subset of natural numbers contains its least element.

We will use the above Axiom to prove the weak form of the principle of mathematical induction. The proof is based on contradiction. That is, suppose that we need to prove that "whenever the statement P holds true, the statement Q holds true as well". A proof by contradiction starts with the assumption that "the statement P holds true and the statement Q does not hold true" and tries to arrive at a contradiction to the validity of the statement P being true.

Theorem: (Principle of Mathematical Induction: Weak Form). Let P (n) be a statement about a positive integer n such that

1. P (1) is true, and

2. P (k + 1) is true whenever one assumes that P (k) is true.

Then P (n) is true for all positive integer n.

Proof. On the contrary, assume that there exists $n_0 \in \mathbb{N}$ such that P (n_0) is not true. Now, consider the set

$$S = \{m \in \mathbb{N} : P(m) \text{ is false}\}.$$

As $n_0 \in S$, $S \neq \emptyset$. So, by Well-Ordering Principle, S must have a least element, say N . By assumption, $N \neq 1$ as P (1) is true. Thus, N $\geq$ 2 and hence $N - 1 \in \mathbb{N}$.

Therefore, from the assumption that N is the least element in S and S contains all those $m \in \mathbb{N}$ for which P (m) is false, one deduces that P (N – 1) holds true as N – 1 < N ≤ 2. Thus, the implication "P (N – 1) is true" and Hypothesis 2 imply that P (N) is true.

This leads to a contradiction and hence our first assumption that there exists $n_0 \in \mathbb{N}$, such that P (n_0) is not true is false.

Example: 1. Prove that $1 + 2 + \cdots + n = \dfrac{n(n+1)}{2}$.

Solution: Verify that the result is true for n = 1. Hence, let the result be true for n. Let us now prove it for n + 1. That is, one needs to show that $1 + 2 + \cdots + n + (n+1) = \dfrac{(n+1)(n+2)}{2}$.

Using Hypothesis 2,

$$1 + 2 + \cdots + n + (n+1) = \frac{n(n+1)}{2} + (n+1) = \frac{n+1}{2}(n+2).$$

Thus, by the principle of mathematical induction, the result follows.

2. Prove that $1^2 + 2^2 + \cdots + n^2 = \dfrac{n(n+1)(2n+1)}{6}$.

Solution: The result is clearly true for n = 1. Hence, let the result be true for n and one needs to show that $1^2 + 2^2 + \cdots + n^2 + (n+1)^2 = \dfrac{(n+1)(n+2)(2(n+1)+1)}{6}$.

Using Hypothesis 2,

$$1^2 + 2^2 + \cdots + n^2 + (n+1)^2 = \frac{n(n+1)(2n+1)}{6} + (n+1)^2$$

$$= \frac{n+1}{6}(n(2n+1) + 6(n+1))$$

$$= \frac{n+1}{6}(2n^2 + 7n + 6) = \frac{(n+1)(n+2)(2n+3)}{6}.$$

Thus, by the principle of mathematical induction, the result follows.

3. Prove that for any positive integer n, $1 + 3 + \cdots + (2n - 1) = n^2$.

Solution: The result is clearly true for n = 1. Let the result be true for n. That is,

$1 + 3 + \cdots + (2n - 1) = n^2$. Now, that

$1 + 3 + \cdots + (2n - 1) + (2n + 1) = n^2 + (2n + 1) = (n + 1)^2$.

Thus, by the principle of mathematical induction, the result follows.

4. AM-GM Inequality: Let $n \in \mathbb{N}$ and suppose we are given real numbers $a_1 \geq a_2 \geq \cdots \geq a_n \geq 0$. Then

$$Arithmetic\ Mean\ (AM) := \frac{a_1 + a_2 + \cdots + a_n}{n} \geq \sqrt[n]{a_1 \cdot a_2 \cdots \cdot a_n} =: (GM)\ Geometric\ Mean.$$

Solution: The result is clearly true for n = 1, 2. So, we assume the result holds for any collection of n non-negative real numbers. Need to prove AM ≥ GM, for any collection of non-negative integers $a_1 \geq a_2 \geq \cdots \geq a_n \geq a_{n+1} \geq 0$.

So, let us assume that $A = \frac{a_1 + a_2 + \ldots + a_n + a_{n+1}}{n+1}$. Then, it can be easily verified that $a_1 \geq A \geq a_{n+1}$ and hence $a_1 - A, A - a_{n+1} \geq 0$. Thus, $(a_1 - A)(A - a_{n+1}) \geq 0$.

Or equivalently,

$$A(a_1 + a_{n+1} - A) \geq a_1 a_{n+1}. \qquad (1)$$

Now, according to our assumption, the AM-GM inequality holds for any collection of n non-negative numbers. Hence, in particular, for the collection $a_2, a_3, \ldots, a_n, a_1 + a_{n+1} - A$.

That is,

$$AM = \frac{a_2 + \cdots + a_n + (a_1 + a_{n+1} - A)}{n} \geq \sqrt[n]{a_2 \cdots a_n \cdot (a_1 + a_{n+1} - A)} = GM. \qquad (2)$$

But

$\frac{a_2 + a_3 + \cdots + a_n + (a_1 + a_{n+1} - A)}{n} = A$. Thus, by Equation (1) and Equation (2), one has

$$A^{n+1} \geq (a_2 \bullet a_3 \ldots a_n \bullet (a_1 + a_{n+1} - A)) \bullet A \geq (a_2 \bullet a_3 \ldots a_n) a_1 a_{n+1}.$$

Therefore, we see that by the principle of mathematical induction, the result follows.

5. Fix a positive integer n and let A be a set with |A| = n. Then prove that P(A) = 2^n.

Solution: Using example, it follows that the result is true for n = 1. Let the result be true for all subset A, for which |A| = n. We need to prove the result for a set A that contains n + 1 distinct elements, say $a_1, a_2, \ldots, a_{n+1}$.

Let B = $\{a_1, a_2, \ldots, a_n\}$. Then B ⊆ A, |B| = n and by induction hypothesis, |P(B)| = 2^n. Also, $P(B) = \{S \subseteq \{a_1, a_2, \ldots, a_n, a_{n+1}\} : a_{n+1} \notin S\}$. Therefore, it can be easily verified that

$$P(A) = P(B) \cup \{S \cup \{a_{n+1}\} : S \in P(B)\}.$$

Also, note that $P(B) \cap \{S \cup \{a_{n+1}\} : S \in P(B)\} = \emptyset$, as $a_{n+1} \notin S$, for all $S \in P(B)$. Hence, using examples we see that

$$\left|P(A)\right| = \left|P(B)\right| + \left|\{S \cup \{a_{n+1}\} : S \in P(B)\}\right| = \left|P(B)\right| + \left|P(B)\right| = 2^n + 2^n = 2^{n+1}.$$

Thus, the result holds for any set that consists of $n + 1$ distinct elements and hence by the principle of mathematical induction, the result holds for every positive integer n.

We state a corollary of the theorem on page no. 24 without proof. The readers are advised to prove it for the sake of clarity.

Corollary (Principle of Mathematical Induction). Let P (n) be a statement about a positive integer n such that for some fixed positive integer n_o,

1. P (n_o) is true,

2. P (k + 1) is true whenever one assumes that P (k) is true.

Then P (n) is true for all positive integer $n \geq n_o$.

Strong Form of the Principle of Mathematical Induction

We are now ready to prove the strong form of the principle of mathematical induction.

Theorem: (Principle of Mathematical Induction: Strong Form). Let P (n) be a statement about a positive integer n such that

1. P (1) is true, and

2. P (k + 1) is true whenever one assume that P (m) is true, for all m, $1 \leq m \leq k$.

Then, P (n) is true for all positive integer n.

Proof. Let R(n) be the statement that "the statement P (m) holds, for all positive integers m with $1 \leq m \leq n$". We prove that R(n) holds, for all positive integers n, using the weak-form of mathematical induction. This will give us the required result as the statement "R(n) holds true" clearly implies that "P (n) also holds true".

As the first step of the induction hypothesis, we see that R(1) holds true (already assumed in the hypothesis of the theorem). So, let us assume that R(n) holds true. We need to prove that R(n + 1)holds true.

The assumption that R(n) holds true is equivalent to the statement "P (m) holds true, for all m, $1 \leq m \leq n$". Therefore, by Hypothesis 2, P (n + 1) holds true. That is, the statements "R(n) holds true" and "P (n + 1) holds true" are equivalent to the statement "P (m) holds true, for all m, $1 \leq m \leq n + 1$". Hence, we have shown that R(n + 1) holds true. Therefore, we see that the result follows, using the weak-form of the principle of mathematical induction.

We state a corollary of the above theorem without proof.

Corollary (Principle of Mathematical Induction). Let P (n) be a statement about a positive integer n such that for some fixed positive integer n_0,

1. $P(n_0)$ is true,

2. P (k + 1) is true whenever one assume that P (m) is true, for all m, $n_0 \le m \le k$.

Then P (n) is true for all positive integer $n \ge n_0$.

Remark (Pitfalls). Find the error in the following arguments:

1. If a set of n balls contains a green ball then all the balls in the set are green.

Solution: If n = 1, we are done. So, let the result be true for any collection of n balls in which there is at least one green ball.

So, let us assume that we have a collection of n + 1 balls that contains at least one green ball. From this collection, pick a collection of n balls that contains at least one green ball. Then by the induction hypothesis, this collection of n balls has all green balls.

Now, remove one ball from this collection and put the ball which was left out. Observe that the ball removed is green as by induction hypothesis all balls were green. Again, the new collection of n balls has at least one green ball and hence, by induction hypothesis, all the balls in this new collection are also green. Therefore, we see that all the n + 1 balls are green. Hence the result follows by induction hypothesis.

2. In any collection of n lines in a plane, no two of which are parallel, all the lines pass through a common point.

Solution: If n = 1, 2 then the result is easily seen to be true. So, let the result be true for any collection of n lines, no two of which are parallel. That is, we assume that if we are given any collection of n lines which are pairwise non-parallel then they pass through a common point.

Now, let us consider a collection of n + 1 lines in the plane. We are also given that no two lines in this collection are parallel. Let us denote these lines by $\ell_1, \ell_2, \ldots, \ell_{n+1}$. From this collection of lines, let us choose the subset $\ell_1, \ell_2, \ldots, \ell_n$, consisting of n lines. By induction hypothesis, all these lines pass through a common point, say P, the point of intersection of the lines ℓ_1 and ℓ_2. Now, consider the collection $\ell_1, \ell_2, \ldots, \ell_{n-1}, \ell_{n+1}$. This collection again consists of n non-parallel lines and hence by induction hypothesis, all these lines pass through a common point. This common point is P itself, as P is the point of intersection of the lines ℓ_1 and ℓ_2. Thus, by the principle of mathematical induction the proof of our statement is complete.

3. Consider the polynomial $f(x) = x^2 - x + 41$. Check that for $1 \le n \le 40$, f (n) is a prime number. Does this necessarily imply that f (n) is prime for all positive integers n? Check that $f(41) = 41^2$ and hence f (41) is not a prime. Thus, the validity is being negated using the proof technique "disproving by counter-example".

Fundamental Theorem of Arithmetic

DISQVISITIONES

ARITHMETICAE

AVCTORE

D. CAROLO FRIDERICO GAVSS

LIPSIAE

IN COMMISSIS APVD GEBR. FLEISCHER, JVR.

1801.

The unique factorization theorem was proved by Gauss with his 1801 book *Disquisitiones Arithmeticae*. In this book, Gauss used the fundamental theorem for proving the law of quadratic reciprocity.

In number theory, the fundamental theorem of arithmetic, also called the unique factorization theorem or the unique-prime-factorization theorem, states that every integer greater than 1 either is prime itself or is the product of prime numbers, and that this product is unique, up to the order of the factors. For example,

$$1200 = 2^4 \times 3^1 \times 5^2 = 3 \times 2 \times 2 \times 2 \times 2 \times 5 \times 5 = 5 \times 2 \times 3 \times 2 \times 5 \times 2 \times 2 = \text{etc.}$$

The theorem is stating two things: first, that 1200 *can* be represented as a product of primes, and second, no matter how this is done, there will always be four 2s, one 3, two 5s, and no other primes in the product.

The requirement that the factors be prime is necessary: factorizations containing composite numbers may not be unique (e.g. $12 = 2 \times 6 = 3 \times 4$).

This theorem is one of the main reasons why 1 is not considered a prime number: if 1 were prime, the factorization would not be unique, as, for example, $2 = 2 \times 1 = 2 \times 1 \times 1 = \dots$

History

Book VII, propositions 30, 31 and 32, and Book IX, proposition 14 of Euclid's *Elements* are essentially the statement and proof of the fundamental theorem.

If two numbers by multiplying one another make some number, and any prime number measure the product, it will also measure one of the original numbers.

—Euclid, Elements Book VII, Proposition 30

Proposition 30 is referred to as Euclid's lemma. And it is the key in the proof of the fundamental theorem of arithmetic.

Any composite number is measured by some prime number.

—Euclid, Elements Book VII, Proposition 31

Proposition 31 is proved directly by infinite descent.

Any number either is prime or is measured by some prime number.

—Euclid, Elements Book VII, Proposition 32

Proposition 32 is derived from proposition 31, and prove that the decomposition is possible.

If a number be the least that is measured by prime numbers, it will not be measured by any other prime number except those originally measuring it.

—Euclid, Elements Book IX, Proposition 14

Book IX, proposition 14 is derived from Book VII, proposition 30, and prove partially that the decomposition is unique – a point critically noted by André Weil. Indeed, in this proposition the exponents are all equal to one, so nothing is said for the general case.

Article 16 of Gauss' *Disquisitiones Arithmeticae* is an early modern statement and proof employing modular arithmetic.

Applications

Canonical Representation of a Positive Integer

Every positive integer $n > 1$ can be represented in exactly one way as a product of prime powers:

$$n = p_1^{\alpha_1} p_2^{\alpha_2} \cdots p_k^{\alpha_k} = \prod_{i=1}^{k} p_i^{\alpha_i}$$

where $p_1 < p_2 < ... < p_k$ are primes and the α_i are positive integers. This representation is commonly extended to all positive integers, including one, by the convention that the empty product is equal to 1 (the empty product corresponds to $k = 0$).

This representation is called the canonical representation of n, or the standard form of n.

For example $999 = 3^3 \times 37$, $1000 = 2^3 \times 5^3$, $1001 = 7 \times 11 \times 13$

Note that factors $p^0 = 1$ may be inserted without changing the value of n (e.g. $1000 = 2^3 \times 3^0 \times 5^3$). In fact, any positive integer can be uniquely represented as an infinite product taken over all the positive prime numbers,

$$n = 2^{n_1} 3^{n_2} 5^{n_3} 7^{n_4} \cdots = \prod p_i^{n_i},$$

where a finite number of the n_i are positive integers, and the rest are zero. Allowing negative exponents provides a canonical form for positive rational numbers.

Arithmetic Operations

The canonical representation, when it is known, allows for a simple representation of the products, gcd, and lcm of two numbers:

$$a \cdot b = 2^{a_2+b_2} 3^{a_3+b_3} 5^{a_5+b_5} 7^{a_7+b_7} \cdots = \prod p_i^{a_{p_i}+b_{p_i}},$$

$$\gcd(a,b) = 2^{\min(a_2,b_2)} 3^{\min(a_3,b_3)} 5^{\min(a_5,b_5)} 7^{\min(a_7,b_7)} \cdots = \prod p_i^{\min(a_{p_i},b_{p_i})},$$

$$\mathrm{lcm}(a,b) = 2^{\max(a_2,b_2)} 3^{\max(a_3,b_3)} 5^{\max(a_5,b_5)} 7^{\max(a_7,b_7)} \cdots = \prod p_i^{\max(a_{p_i},b_{p_i})}.$$

However, as Integer factorization of large integers is much harder than computing their product, gcd or lcm, these formulas have, in practice, a limited usage.

Arithmetical Functions

Many arithmetical functions are defined using the canonical representation. In particular, the values of additive and multiplicative functions are determined by their values on the powers of prime numbers.

Proof

The proof uses Euclid's lemma (*Elements* VII, 30): if a prime p divides the product of two natural numbers a and b, then p divides a or p divides b.

Existence

We need to show that every integer greater than 1 is either prime or a product of primes. For the base case, note that 2 is prime. By induction: assume true for all numbers between 1 and n. If n is prime, there is nothing more to prove. Otherwise, there are integers a and b, where $n = ab$ and $1 < a \le b < n$. By the induction hypothesis, $a = p_1 p_2 ... p_j$ and $b = q_1 q_2 ... q_k$ are products of primes. But then $n = ab = p_1 p_2 ... p_j q_1 q_2 ... q_k$ is a product of primes.

Uniqueness

Assume that $s > 1$ is the product of prime numbers in two different ways:

$$s = p_1 p_2 \cdots p_m$$
$$= q_1 q_2 \cdots q_n.$$

We must show $m = n$ and that the q_j are a rearrangement of the p_i.

By Euclid's lemma, p_1 must divide one of the q_j; relabeling the q_j if necessary, say that p_1 divides q_1. But q_1 is prime, so its only divisors are itself and 1. Therefore, $p_1 = q_1$, so that

$$\frac{s}{p_1} = p_2 \cdots p_m$$
$$= q_2 \cdots q_n.$$

Reasoning the same way, p_2 must equal one of the remaining q_j. Relabeling again if necessary, say $p_2 = q_2$. Then

$$\frac{s}{p_1 p_2} = p_3 \cdots p_m$$
$$= q_3 \cdots q_n.$$

This can be done for each of the m p_i's, showing that $m \leq n$ and every p_i is a q_j. Applying the same argument with the p's and q's reversed shows $n \leq m$ (hence $m = n$) and every q_j is a p_i.

Elementary Proof of Uniqueness

The fundamental theorem of arithmetic can also be proved without using Euclid's lemma, as follows:

Assume that $s > 1$ is the smallest positive integer which is the product of prime numbers in two different ways. If s were prime then it would factor uniquely as itself, so there must be at least two primes in each factorization of s:

$$s = p_1 p_2 \cdots p_m$$
$$= q_1 q_2 \cdots q_n.$$

If any $p_i = q_j$ then, by cancellation, $s/p_i = s/q_j$ would be another positive integer, different from s, which is greater than 1 and also has two distinct factorizations. But s/p_i is smaller than s, meaning s would not actually be the smallest such integer. Therefore every p_i must be distinct from every q_j.

Without loss of generality, take $p_1 < q_1$ (if this is not already the case, switch the p and q designations.) Consider

$$t = (q_1 - p_1)(q_2 \cdots q_n),$$

and note that $1 < q_2 \leq t < s$. Therefore t must have a unique prime factorization. By rearrangement we see,

$$t = q_1(q_2 \cdots q_n) - p_1(q_2 \cdots q_n)$$
$$= s - p_1(q_2 \cdots q_n)$$
$$= p_1((p_2 \cdots p_m) - (q_2 \cdots q_n)).$$

Here $u = ((p_2 \cdots p_m) - (q_2 \cdots q_n))$ is positive, for if it were negative or zero then so would be its product with p_1, but that product equals t which is positive. So u is either 1 or factors into primes. In either case, $t = p_1 u$ yields a prime factorization of t, which we know to be unique, so p_1 appears in the prime factorization of t.

If $(q_1 - p_1)$ equaled 1 then the prime factorization of t would be all q's, which would preclude p_1 from appearing. Thus $(q_1 - p_1)$ is not 1, but is positive, so it factors into primes: $(q_1 - p_1) = (r_1 \ldots r_h)$. This yields a prime factorization of

$$t = (r_1 \cdots r_h)(q_2 \cdots q_n),$$

which we know is unique. Now, p_1 appears in the prime factorization of t, and it is not equal to any q, so it must be one of the r's. That means p_1 is a factor of $(q_1 - p_1)$, so there exists a positive integer k such that $p_1 k = (q_1 - p_1)$, and therefore

$$p_1(k+1) = q_1.$$

But that means q_1 has a proper factorization, so it is not a prime number. This contradiction shows that s does not actually have two different prime factorizations. As a result, there is no smallest positive integer with multiple prime factorizations, hence all positive integers greater than 1 factor uniquely into primes.

Generalizations

The first generalization of the theorem is found in Gauss's second monograph (1832) on biquadratic reciprocity. This paper introduced what is now called the ring of Gaussian integers, the set of all complex numbers $a + bi$ where a and b are integers. It is now denoted by $\mathbb{Z}[i]$. He showed that this ring has the four units ± 1 and $\pm i$, that the non-zero, non-unit numbers fall into two classes, primes and composites, and that (except for order), the composites have unique factorization as a product of primes.

Similarly, in 1844 while working on cubic reciprocity, Eisenstein introduced the ring $\mathbb{Z}[\omega]$, where $\omega = \dfrac{-1+\sqrt{-3}}{2}$, $\omega^3 = 1$ is a cube root of unity. This is the ring of Eisenstein integers, and he proved it has the six units $\pm 1, \pm \omega, \pm \omega^2$ and that it has unique factorization.

However, it was also discovered that unique factorization does not always hold. An example is given by $\mathbb{Z}[\sqrt{-5}]$. In this ring one has

$$6 = 2 \cdot 3 = (1 + \sqrt{-5})(1 - \sqrt{-5}).$$

Examples like this caused the notion of "prime" to be modified. In $\mathbb{Z}[\sqrt{-5}]$ it can be proven that if any of the factors above can be represented as a product, e.g. $2 = ab$, then one of a or b must be a unit. This is the traditional definition of "prime". It can also be proven that none of these factors obeys Euclid's lemma; e.g. 2 divides neither $(1 + \sqrt{-5})$ nor $(1 - \sqrt{-5})$ even though it divides their product 6. In algebraic number theory 2 is called irreducible in $\mathbb{Z}[\sqrt{-5}]$ (only divisible by itself or a unit) but not prime in $\mathbb{Z}[\sqrt{-5}]$ (if it divides a product it must divide one of the factors). The mention of $\mathbb{Z}[\sqrt{-5}]$ is required because 2 is prime and irreducible in $\mathbb{Z}$. Using these definitions it can be proven that in any ring a prime must be irreducible. Euclid's classical lemma can be rephrased as "in the ring of integers $\mathbb{Z}$ every irreducible is prime". This is also true in $\mathbb{Z}[i]$ and $\mathbb{Z}[\omega]$ but not in $\mathbb{Z}[\sqrt{-5}]$.

The rings in which factorization into irreducibles is essentially unique are called unique factorization domains. Important examples are polynomial rings over the integers or over a field, Euclidean domains and principal ideal domains.

In 1843 Kummer introduced the concept of ideal number, which was developed further by Dedekind (1876) into the modern theory of ideals, special subsets of rings. Multiplication is defined for ideals, and the rings in which they have unique factorization are called Dedekind domains.

There is a version of unique factorization for ordinals, though it requires some additional conditions to ensure uniqueness.

Division Algorithm and the Fundamental Theorem of Arith- metic

In the next few pages, we will try to study properties of integers that will be required later. We start with a lemma, commonly known as the "division algorithm". The proof again uses the technique "proof by contradiction".

Lemma (Division Algorithm) Let a and b be two integers with b > 0. Then there exist unique integers q, r such that a = qb + r, where $0 \le r < b$. The integer q is called the quotient and r, the remainder.

Proof. Without loss of generality, assume that a ≥ 0 and consider the set $S = \{a + bx : x \in \mathbb{Z}\} \cap \mathbb{N}$. Clearly, a ∈ S and hence S is a non-empty subset of $\mathbb{N}$. Therefore, by Well-Ordering Principle, S contains its least element, say s_0. That is, there exists $x_0 \in \mathbb{Z}$, such that $s_0 = a + bx_0$. We claim that $0 \le s_0 < b$.

As $s_0 \in S \subset \mathbb{N}$, one has $s_0 \ge 0$. So, let if possible assume that $s_0 \ge b$. This implies that $s_0 - b \ge 0$ and hence $s_0 - b = a + b(x_0 - 1) \in S$, a contradiction to the assumption that s_0 was the least element of S. Hence, we have shown the existence of integers q, r such that a = qb + r with $0 \le r < b$.

Uniqueness: Let if possible q_1, q_2, r_1 and r_2 be integers with $a = q_1 b + r_1 = q_2 b + r_2$, with $0 \le r_1 \le r_2 < b$. Therefore, $r_2 - r_2 \ge 0$ and thus, $0 \le (q_1 - q_2)b = r_2 - r_1 < b$. Hence, we have obtained a multiple of b that is strictly less than b. But this can happen only if the multiple is

1. Multiple is 0. That is, $0 = (q_1 - q_2)b = r_2 - r_1$. Thus, one obtains $r_1 = r_2$ and $q_1 = q_2$ and the proof of uniqueness is complete.

This completes the proof of the lemma.

Definition (Greatest Common Divisor): 1. An integer a is said to divide an integer b, denoted a|b, if b = ac, for some integer c. Note that c can be a negative integer.

2. Greatest Common Divisor: Let $a, b \in \mathbb{Z} \setminus \{0\}$. Then the greatest common divisor of a and b, denoted gcd(a, b), is the largest positive integer c such that

   (a) c divides a and b, and

   (b) if d is any positive integer dividing a and b, then d divides c as well.

3. Relatively Prime/Coprime Integers: Two integers a and b are said to be relatively prime if gcd(a, b) = 1.

Theorem (Euclid's Algorithm). Let a and b be two non-zero integers. Then there exists an integer d such that

1. d = gcd(a, b), and

2. There exist integers x_0, y_0 such that $d = ax_0 + by_0$.

Proof. Consider the set $S = \{ax + by : x, y \in \mathbb{Z}\} \cap \mathbb{N}$. Then, either a $\in$ S or $-$a $\in$ S, as exactly one of them is an element of $\mathbb{N}$ and both a = a • 1 + b • 0 and $-$a = a • ($-$1) + b • 0 are elements of the set $\{ax + by : x, y \in \mathbb{Z}\}$. Thus, S is non-empty subset of $\mathbb{N}$. So, by Well-Ordering Principle, S contains its least element, say d. As d $\in$ S, there exist integers x_0, y_0 such that $d = ax_0 + by_0$.

We claim that d obtained as the least element of S also equals gcd(a, b). That is, we need to show that d satisfies both the conditions of Definition.

We first show that d|a. By division algorithm, there exist integers q and r such that a = dq+r, with 0 $\leq$ r < d. Thus, we need to show that r = 0. On the contrary, assume that $r \neq 0$. That is, 0 < r < d. Then by definition, $r \in \mathbb{N}$ and $r = a - dq = a - q \cdot (ax_0 + by_0) = a \cdot (1 - qx_0) + b \cdot (-qy_0) \in \{ax + by : x, y \in \mathbb{Z}\}$. Hence, r $\in$ S and by our assumption r < d. This contradicts the fact that d was the least element of S. Thus, our assumption that $r \neq 0$ is false and hence a = dq. This implies that d|a. In a similar way, it can be shown that d|b.

Now, assume that there is an integer c such that c divides both a and b. We need to show that c|d. Observe that as c divides both a and b, c also divides both ax_0 and by_0 and hence c also divides $ax_0 + by_0 = d$. Thus, we have shown that d satisfies both the conditions of Definition and therefore, the proof of the theorem is complete.

The above theorem is often stated as "the gcd(a, b) is a linear combination of the numbers a and b".

Example 1. Consider two integers, say 155 and $-$275. Then, by division algorithm, one obtains

$$-275 = (-2) \cdot 155 + 35 \qquad 155 = 4 \cdot 35 + 15$$

$$35 = 2 \cdot 15 + 5 \qquad 15 = 3 \cdot 5.$$

Hence, 5 = gcd(155, $-$275) and 5 = 9 • ($-$275) + 16 • 155, as

5 = 35$-$2•15 = 35$-$2(155$-$4•35) = 9•35$-$2•155 = 9($-$275+2•155)$-$2•155 = 9•($-$275)+16•155.

Also, note that 275 = 5•55 and 155 = 5•31 and thus, 5 = (9+31x)•($-$275)+(16+55x)•155, for all $x \in \mathbb{Z}$. Therefore, we see that there are infinite number of choices for the pair $(x\ y) \in \mathbb{Z}$, for which $d = ax + by$.

2. In general, given two non-zero integers a and b, we can use the division algorithm to get gcd(a, b). This algorithm is also attributed to Euclid. Without loss of generality, assume that both a and b are positive and a > b. Then the algorithm proceeds as follows:

$$a \;=\; bq_0 + r_0 \quad with \; 0 \le r_0 < b, \qquad\qquad b = r_0 q_1 + r_1$$

$$r_0 \;=\; r_1 q_2 + r_2 \quad with \; 0 \le r_2 < r_1, \qquad\qquad r_1 = r_2 q_3 + r_3$$

$$\vdots \;=\; \vdots$$

$$r_{\ell-1} \;=\; r_\ell q_{\ell+1} + r_{\ell+1} \quad with \; 0 \le r_{\ell+1} < r_\ell, \qquad r_\ell = r_{\ell+1} q_{\ell+2}.$$

The process will take at most $b - 1$ steps as $0 \le r_0 < b$. Also, note that $\gcd(a, b) = r_{\ell+1}$ and it can be recursively obtained, using backtracking. That is,

$$r_{\ell+1} = r_{\ell-1} - r_\ell q_{\ell+1} = r_{\ell-1} - q_{\ell+1}\left(r_{\ell-2} - r_{\ell-1}q_\ell\right) = r_{\ell-1}\left(1 + q_{\ell+1}q_\ell\right) - q_{\ell+1}r_{\ell-2} = \cdots.$$

To proceed further, we need the following definitions:

Definition (Prime/Composite Numbers):

1. The positive integer 1 is called the unity or the unit element of $\mathbb{Z}$.

2. A positive integer p is said to be a prime, if p has exactly two factors, namely, 1 and p itself.

3. An integer r is called composite if $\ne$, -1 and is not a prime.

We are now ready to prove an important result that helps us in proving the fundamental theorem of arithmetic.

Lemma (Euclid's Lemma). Let p be a prime and let a, $b \in \mathbb{Z}$. If $p|ab$ then either $p|a$ or $p|b$.

Proof. If $p|a$, then we are done. So, let us assume that p does not divide a. But p is a prime and hence $\gcd(p, a) = 1$. Thus, by Euclid's algorithm, there exist integers x, y such that $1 = ax + py$. Therefore,

$$b = b \cdot 1 = b \cdot (ax + py) = ab \cdot x + p \cdot by.$$

Now, the condition $p|ab$ implies that p divides $ab \cdot x + p \cdot by = b$. Thus, we have shown that if $p|ab$ then either $p|a$ or $p|b$.

Now, we are ready to prove the fundamental theorem of arithmetic that states that "every positive integer greater than 1 is either a prime or is a product of primes. This product is unique, except for the order in which the prime factors appear".

Theorem (Fundamental Theorem of Arithmetic). Let $n \in \mathbb{N}$ with $n \ge 2$. Then there exist prime numbers $p_1 > p_2 > \cdots > p_k$ and positive integers $s_1, s_2, \cdots, s_k$ such that $n = p_1^{s_1} p_2^{s_2} \cdots p_k^{s_k}$, for some $k \ge 1$. Moreover, if n also equals $q_1^{t_1} q_2^{t_2} \cdots q_l^{t_l}$, for distinct primes $q_1, q_2, \cdots, q_l$ and positive integers $t_1, t_2, \cdots, t_l$ then $k = \ell$ and for each i, $1 \le i \le k$, there exists j, $1 \le j \le k$ such that $p_i = q_j$ and $s_i = t_j$.

Proof. We prove the result using the strong form of the principle of mathematical induction. If n equals a prime, say p then clearly $n = p^1$ and hence the first step of the induction holds true. Hence, let us assume that the result holds for all positive integers that are less than n. We need to prove the result for the positive integer n.

If n itself is a prime then we are done. Else, there exists positive integers a and b such that n = ab and $1 \le a, b < n$. Thus, by the strong form of the induction hypothesis, there exist primes p_i's, q_j's and positive integers s_i and t_j's such that $a = p_1^{s_1} p_2^{s_2} \cdots p_k^{s_k}$, for some $k \ge 1$ and $b = q_1^{t_1} q_2^{t_2} \cdots q_l^{t_l}$, for some ℓ. Hence,

$$n = ab = p_1^{s_1} p_2^{s_2} \cdots p_k^{s_k} q_1^{t_1} q_2^{t_2} \cdots q_l^{t_l}.$$

Now, if some of the p_i's and q_j's are equal, they can be multiplied together to obtain n as a product of distinct prime powers.

Thus, using the strong form of the principle of mathematical induction, the result is true for all positive integer n. As far as the uniqueness is concerned, it follows by a repeated application of Lemma.

To see this, observe that p_1 divides $n = q_1^{t_1} q_2^{t_2} \cdots q_l^{t_l}$ implies that p_1 divides exactly one of them (the primes are distinct), say q_1. Also, it is clear that in this case $s_1 = t_1$. For otherwise, either p_1 will divide $q_2^{t_2} \cdots q_l^{t_l}$, or $q_1 = p_1$ will divide $p_2^{s_2} \cdots p_k^{s_k}$. This process can be continued a finite number of times to get the required result.

As an application of the fundamental theorem of arithmetic, one has the following well known result. This is the first instance where we have used the contrapositive argument technique to prove the result.

Corollary Let $n \in \mathbb{N}$ with $n \ge 2$. Suppose that for any prime $p \le \sqrt{n}$, p does not divide n then n is prime.

Proof. Suppose n is not a prime. Then there exists positive integers a and b such that n = ab and $2 \le a, b \le n$. Also, note that at least one of them, say $a \le \sqrt{n}$. For if, both $a, b > \sqrt{n}$ then n = ab > n, giving us a contradiction.

Since $a \le \sqrt{n}$, by the theorem, one of its prime factors, say p will satisfy $p \le a \le \sqrt{n}$. Thus, if n has no prime divisor less than or equal to $\sqrt{n}$ then n must be itself be a prime.

Relations, Partitions and Equivalence Relation

We start with the definition of cartesian product of two sets and to define relations.

Definition (Cartesian Product). Let A and B be two sets. Then their cartesian product, denoted A × B, is defined as A × B = {(a, b) : a ∈ A, b ∈ B}.

Example

1. Let A = {a, b, c} and B = {1, 2, 3, 4}. Then

A × A = {(a, a), (a, b), (a, c), (b, a), (b, b), (b, c), (c, a), (c, b), (c, c)}.

A × B = {(a, 1), (a, 2), (a, 3), (a, 4), (b, 1), (b, 2), (b, 3), (b, 4), (c, 1), (c, 2), (c, 3), (c, 4)}.

2. The Euclidean plane, denoted $\mathbb{R}^2 = \mathbb{R} \times \mathbb{R} = \{(x, y) : x \in \mathbb{R}\}$.

Definition (Relation) A relation R on a non-empty set A, is a subset of A × A.

Example

1. Let A = {a, b, c, d}. Then, some of the relations R on A are:

    (a) R = A × A.

    (b) R = {(a, a), (b, b), (c, c), (d, d), (a, b), (a, c), (b, c)}.

    (c) R = {(a, a), (b, b), (c, c)}.

    (d) R = {(a, a), (a, b), (b, a), (b, b), (c, d)}.

    (e) R = {(a, a), (a, b), (b, a), (a, c), (c, a), (c, c), (b, b)}.

    (f) R = {(a, b), (b, c), (a, c), (d, d)}.

2. Consider the set $\mathbb{Z}^* = \mathbb{Z}\setminus\{0\}$. Some of the relations on $\mathbb{Z}^*$ are as follows:

    (a) $R = \{(a,b) \in \mathbb{Z}^* \times \mathbb{Z}^* : a \mid b\}$.

    (b) Fix a positive integer n and define $R = \{(a,b) \in \mathbb{Z}^2 : n \text{ divides } a - b\}$.

    (c) $R = \{(a,b) \in \mathbb{Z}^2 : a \leq b\}$.

    (d) $R = \{(a,b) \in \mathbb{Z}^2 : a > b\}$.

3. Consider the set $\mathbb{R}^2$. Also, let us write $x = (x_1, x_2)$ and $y = (y_1, y_2)$. Then some of the relations on $\mathbb{R}^2$ are as follows:

    (a) $R = \{(x,y) \in \mathbb{R}^2 \times \mathbb{R}^2 : |x|^2 = x_1^2 + x_2^2 = y_1^2 + y_2^2 = |y|^2\}$.

    (b) $R = \{(x, y) \in \mathbb{R}^2 \times \mathbb{R}^2 : x = \alpha y \text{ for some } \alpha \in \mathbb{R}^*\}$.

    (c) $R = \{(x,y) \in \mathbb{R}^2 \times \mathbb{R}^2 : 4x_1^2 + 9x_2^2 = 4y_1^2 + 9y_2^2\}$.

    (d) $R = \{(x, y) \in \mathbb{R}^2 \times \mathbb{R}^2 : x - y = \alpha(1,1) \text{ for some } \alpha \in \mathbb{R}^*\}$.

    (e) Fix a $c \in \mathbb{R}$. Now, define $R = \{(x,y) \in \mathbb{R}^2 \times \mathbb{R}^2 : y_2 - x_2 = c(y_1 - x_1)\}$.

    (f) $R = \{(x, y) \in \mathbb{R}^2 \times \mathbb{R}^2 : |x| = \alpha |y|\}$, for some positive real number α.

4. Let A be the set of triangles in the plane. Then R = {(a, b) ∈ A² : a ~ b}, where ~ stands for similarity of triangles.

5. In $\mathbb{R}$, define a relation $R = \{(a,b) \in \mathbb{R}^2 : |a - b| \text{ is an integer}\}$.

6. Let A be any non-empty set and consider the set P(A). Then one can define a relation R on P(A) by R = {(S, T) ∈ P(A) × P(A) : S ⊂ T }.

Now that we have seen quite a few examples of relations, let us look at some of the properties that are of interest in mathematics.

Definition: Let R be a relation on a non-empty set A. Then R is said to be

1. reflexive if $(a, a) \in R$, for all $a \in A$.

2. symmetric if $(b, a) \in R$ whenever $(a, b) \in R$.

3. anti-symmetric if, for all $a, b \in A$, the conditions (a, b), $(b, a) \in R$ implies that $a = b$ in A.

4. transitive if, for all $a, b, c \in A$, the conditions (a, b), $(b, c) \in R$ implies that $(a, c) \in R$.

We are now ready to define a relation that appears quite frequently in mathematics. Before doing so, let us either use the symbol $\sim$ or $\underset{\sim}{R}$ for relation. That is, if $a, b \in A$ then $a \underset{\sim}{R} b$ or $\sim b$ will stand for $(a, b) \in R$.

Definition:. Let $\sim$ be a relation on a non-empty set A. Then $\sim$ is said to form an equivalence relation if $\sim$ is reflexive, symmetric and transitive.

The equivalence class containing $a \in A$, denoted $[a]$, is defined as $[a] := \{b \in A : b \sim a\}$.

Example:

1. Let $a, b \in \mathbb{Z}$. Then $a \sim b$, if 10 divides $a - b$. Then verify that $\sim$ is an equivalence relation. Moreover, the equivalence classes can be taken as $[0]$, $[1]$, $\ldots$, $[9]$. Observe that, for $0 \le i \le 9, [i] = \{10n + i : n \in \mathbb{Z}\}$. This equivalence relation in modular arithmetic is written as $a \equiv b$ (mod 10).

In general, for any fixed positive integer n, the statement "$a \equiv b$ (mod n)" (read "a is equivalent to b modulo n") is equivalent to saying that $a \sim b$ if n divides $a - b$.

2. Determine the equivalence relations that appear in the first set of examples given on page no. 38. Also, for each equivalence relation, determine a set of equivalence classes.

Definition: (Partition of a set). Let A be a non-empty set. Then a partition Π of A, into m-parts, is a collection of non-empty subsets $A_1, A_2, \ldots, A_m$, of A, such that

1. $A_i \cap A_j = \emptyset$ (empty set), for $1 \le i \ne j \le m$ and

2. $\displaystyle\bigcup_{i=1}^{m} Ai = A$.

Example: 1. The partitions of $A = \{a, b, c, d\}$ into

(a) 3-parts are $a|\,b|\,cd$, $\;a|\,bc|\,d$, $\;ac|\,b|\,d$, $\;a|\,bd|\,c$, $\;ad|\,b|\,c$, $\;ab|\,c|\,d$, where the expression $a|bc|d$ represents the partition $A_1 = \{a\}$, $A_2 = \{b, c\}$ and $A_3 = \{d\}$.

(b) 2-parts are $a|\,bcd$, $b|\,acd$, $c|\,abd$, $\;d|\,abc$, $\;ab|\,cd$, $\;ac|\,bd$ and $\;ad|\,bc$.

2. Let $A = \mathbb{Z}$ and define

(a) $A_0 = \{2x : x \in \mathbb{Z}\}$ and $A_1 = \{2x+1 : x \in \mathbb{Z}\}$. Then $\Pi = \{A_0, A_1\}$ forms a partition of $\mathbb{Z}$ into odd and even integers.

(b) $A_i = \{10n + i : n \in \mathbb{Z}\}$, for $i = 1, 2, \ldots, 10$. Then $\Pi = \{A_1, A_2, \ldots, A_{10}\}$ forms a partition of $\mathbb{Z}$.

3. $A_1 = \{0,1\}$, $A_2 = \{n \in \mathbb{N} : n \text{ is a prime}\}$ and $A_3 = \{n \in \mathbb{N} : n \geq 3, \ n \text{ is composite}\}$. Then $\Pi = \{A_1, A_2, A_3\}$ is a partition of $\mathbb{N}$.

4. Let A = {a, b, c, d}. Then Π = {{a}, {b, d}, {c}} is a partition of A.

Observe that the equivalence classes produced indeed correspond to the non-empty sets A_i's, defined. In general, such a statement is always true. That is, suppose that A is a non-empty set with an equivalence relation ~. Then the set of distinct equivalence classes of ~ in A, gives rise to a partition of A. Conversely, given any partition Π of A, there is an equivalence relation on A whose distinct equivalence classes are the elements of Π. This is proved as the next result.

Theorem: Let A be a non-empty set.

1. Also, let ~ define an equivalence relation on the set A. Then the set of distinct equivalence classes of ~ in A gives a partition of A.

2. Let I be a non-empty index set such that {A_i : i $\in$ I} gives a partition of A. Then there exists an equivalence relation on A whose distinct equivalence classes are exactly the sets A_i, i $\in$ I.

Proof. Since ~ is reflexive, a ~ a, for all a $\in$ A. Hence, the equivalence class [a] contains a, for each a $\in$ A. Thus, the equivalence classes are non-empty and clearly, their union is the whole set A. We need to show that if [a] and [b] are two equivalence classes of ~ then either $[a] = [b]$ or $[a] \cap [b] = \emptyset$.

Let $x \in [a] \cap [b]$. Then by definition, $x \sim a$ and $x \sim b$. Since ~ is symmetric, one also has $a \sim x$. Therefore, we see that $a \sim x$ and $x \sim b$ and hence, using the transitivity of ~, a ~ b. Thus, by definition, a $\in$ [b] and hence [a] $\subseteq$ [b]. But a ~ b, also implies that b ~ a (~ is symmetric) and hence [b] $\subseteq$ [a]. Thus, we see that if $[a] \cap [b] \neq \emptyset$, then [a] = [b]. This proves the first part of the theorem.

For the second part, define a relation ~ on A as follows: for any two elements a, b $\in$ A, a ~ b if there exists an i, i $\in$ I such that a, b $\in A_i$. It can be easily verified that ~ is indeed reflexive, symmetric and transitive. Also, verify that the equivalence classes of ~ are indeed the sets A_i, i $\in$ I.

A Comprehensive Study of Function

Functions are relations between sets of inputs and outputs where the input is somehow related to the output. The inputs are known as domain, the outputs codomain and the set that denotes their relationality as graph. This chapter elucidates the crucial theories and principles of discrete mathematics.

Function (Mathematics)

In mathematics, a function is a relation between a set of inputs and a set of permissible outputs with the property that each input is related to exactly one output. An example is the function that relates each real number x to its square x^2. The output of a function f corresponding to an input x is denoted by $f(x)$ (read "f of x"). In this example, if the input is -3, then the output is 9, and we may write $f(-3) = 9$. Likewise, if the input is 3, then the output is also 9, and we may write $f(3) = 9$. (The same output may be produced by more than one input, but each input gives only one output.) The input variable(s) are sometimes referred to as the argument(s) of the function.

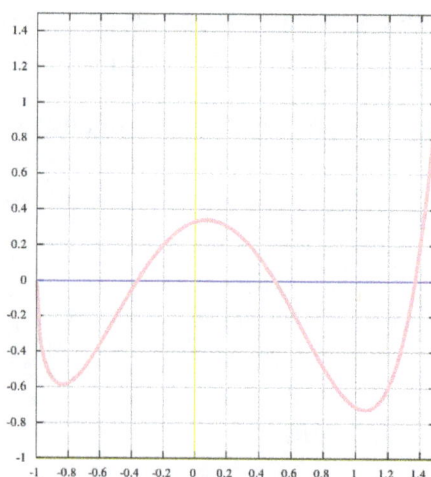

The red curve is the graph of a function f in the Cartesian plane, consisting of all points with coordinates of the form $(x, f(x))$. The property of having one output for each input is represented geometrically by the fact that each vertical line (such as the yellow line through the origin) has exactly one crossing point with the curve.

Functions of various kinds are "the central objects of investigation" in most fields of modern mathematics. There are many ways to describe or represent a function. Some functions may be defined by a formula or algorithm that tells how to compute the output for a given input. Others are given by a picture, called the graph of the function. In science, functions are sometimes defined by a table that gives the outputs for selected inputs. A function could be described implicitly, for example as the inverse to another function or as a solution of a differential equation.

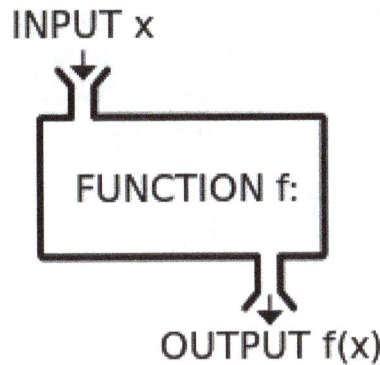

INPUT x

FUNCTION f:

OUTPUT f(x)

A function f takes an input x, and returns a single output $f(x)$. One metaphor describes the function as a "machine" or "black box" that for each input returns a corresponding output.

In modern mathematics, a function is defined by its set of inputs, called the *domain*; a set containing the set of outputs, and possibly additional elements, as members, called its *codomain* (or *target*); and the set of all input-output pairs, called its *graph*. Sometimes the codomain is called the function's "range", but more commonly the word "range" is used to mean, instead, specifically the set of outputs (this is also called the *image* of the function). For example, we could define a function using the rule $f(x) = x^2$ by saying that the domain and codomain are the real numbers, and that the graph consists of all pairs of real numbers (x, x^2). The image of this function is the set of non-negative real numbers. Collections of functions with the same domain and the same codomain are called function spaces, the properties of which are studied in such mathematical disciplines as real analysis, complex analysis, and functional analysis.

In analogy with arithmetic, it is possible to define addition, subtraction, multiplication, and division of functions, in those cases where the output is a number. Another important operation defined on functions is function composition, where the output from one function becomes the input to another function.

Introduction and Examples

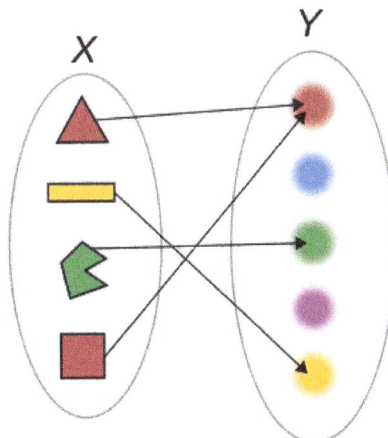

A function that associates any of the four colored shapes to its color.

For an example of a function, let X be the set consisting of four shapes: a red triangle, a yellow rectangle, a green hexagon, and a red square; and let Y be the set consisting of five colors: red, blue, green, pink, and yellow. Linking each shape to its color is a function from X to Y: each shape is linked to a color (i.e., an element in Y), and each shape is "linked", or "mapped", to exactly one color. There is no shape that lacks a color and no shape that has more than one color. This function will be referred to as the "color-of-the-shape function".

The input to a function is called the argument and the output is called the value. The set of all permitted inputs to a given function is called the domain of the function, while the set of permissible outputs is called the codomain. Thus, the domain of the "color-of-the-shape function" is the set of the four shapes, and the codomain consists of the five colors. The concept of a function does *not* require that every possible output is the value of some argument, e.g. the color blue is not the color of any of the four shapes in X.

A second example of a function is the following: the domain is chosen to be the set of natural numbers (1, 2, 3, 4, ...), and the codomain is the set of integers (..., −3, −2, −1, 0, 1, 2, 3, ...). The function associates to any natural number n the number $4-n$. For example, to 1 it associates 3 and to 10 it associates −6.

A third example of a function has the set of polygons as domain and the set of natural numbers as codomain. The function associates a polygon with its number of vertices. For example, a triangle is associated with the number 3, a square with the number 4, and so on.

The term range is sometimes used either for the codomain or for the set of all the actual values a function has.

Definition

In order to avoid the use of the informally defined concepts of "rules" and "associates", the above intuitive explanation of functions is completed with a formal definition. This definition relies on the notion of the Cartesian product. The Cartesian product of two sets X and Y is the set of all ordered pairs, written (x, y), where x is an element of X and y is an element of Y. The x and the y are called the components of the ordered pair. The Cartesian product of X and Y is denoted by $X \times Y$.

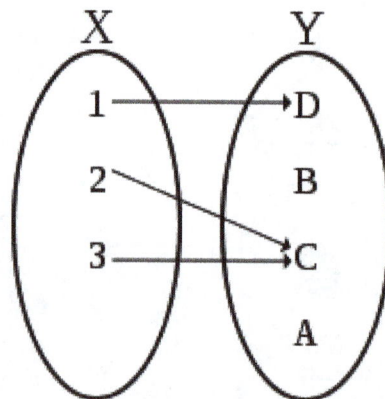

The above diagram represents a function with domain {1, 2, 3}, codomain {A, B, C, D} and set of ordered pairs {(1,D), (2,C), (3,C)}. The image is {C,D}.

A function f from X to Y is a subset of the Cartesian product $X \times Y$ subject to the following condition: every element of X is the first component of one and only one ordered pair in the subset.

In other words, for every x in X there is exactly one element y such that the ordered pair (x, y) is contained in the subset defining the function f. This formal definition is a precise rendition of the idea that to each x is associated an element y of Y, namely the uniquely specified element y with the property just mentioned.

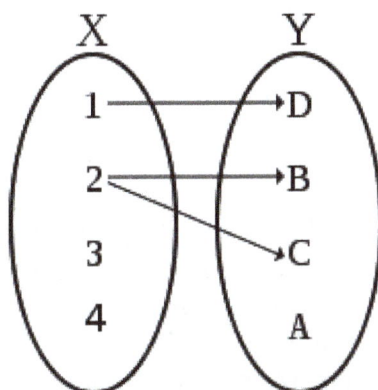

However, this second diagram does *not* represent a function. One reason is that 2 is the first element in more than one ordered pair. In particular, $(2, B)$ and $(2, C)$ are both elements of the set of ordered pairs. Another reason, sufficient by itself, is that 3 is not the first element (input) for any ordered pair. A third reason, likewise, is that 4 is not the first element of any ordered pair.

Considering the "color-of-the-shape" function above, the set X is the domain consisting of the four shapes, while Y is the codomain consisting of five colors. There are twenty possible ordered pairs (four shapes times five colors), one of which is

("yellow rectangle", "red").

The "color-of-the-shape" function described above consists of the set of those ordered pairs,

(shape, color)

where the color is the actual color of the given shape. Thus, the pair ("red triangle", "red") is in the function, but the pair ("yellow rectangle", "red") is not.

Notation

A function f is commonly declared by stating its domain X and codomain Y using the expression

$$f : X \to Y$$

or

$$X \xrightarrow{f} Y.$$

In this context, the elements of X are called arguments of f. For each argument x, the corresponding unique y in the codomain is called the function value at x or the *image* of x under f. It is written as $f(x)$. One says that f associates y with x or maps x to y. This is abbreviated by

$$\ddot{u} = \quad \ddot{u}$$

A general function, to be defined for a particular context, is usually denoted by a single letter, most often the lower-case letters f, g, h. Special functions that have universally (or widely) recognized names and definitions most often have 2- to 4-letter symbols. These functions include, for instance, the trigonometric functions sine, cosine, and tangent with symbols sin, cos, tan (*occasionally* tg), the natural logarithm function, denoted by ln or $\log_e$, and the signum function, denoted by sgn. The value of the function f at a point x is denoted $f(x)$, while the value of a function like sine at x can be notated as $\sin(x)$ or $\sin x$, with the parentheses omitted if no ambiguity arises. Note that by convention, general functions are displayed using an italicized letter (like a variable), while special functions are set in roman type.

Although tending to pedantry when carefully observed on every occasion, one should distinguish between a *function* ("the machine") and its *value at a particular point* ("the output"): the symbol f represents the function, while $f(x)$ is its value at x. Thus, it is sloppy and an abuse of notation to say: "*Let $f(x)$ be the function $x^2 + 1$.*" A careful, precise statement should instead be "*Let f : R → R be the function that maps x to the value $x^2 + 1$,*" or "*Let f: R → R be the function (defined by) $x \mapsto x^2 + 1$*". In practice, sloppy statements are made (either intentionally or unintentionally) in order to avoid cumbersome circumlocutions like these.

The distinction between a function and its value becomes important, for instance, when one wishes to talk about the duality between a function and its argument. In these situations, it is desirable to show the symmetry between function and argument and place the function and the argument on an equal footing. One way to do so is to use *bracket notation*: we write $[x, f]$ for the expression $f(x)$. As examples, we can combine the bracket notation with the dot notation (*discussed below*) in the expression $[x_0, \cdot]$ to represent the mapping (an example of a functional) $f \mapsto [x_0, f] = f(x_0)$ for a given, fixed x_0, without having to introduce the letter f, which is merely a placeholder ("dummy"). Although this notation might allow for cleaner expression of abstract mappings, applying it toward functions containing rational or polynomial expressions is awkward and unwieldy. Thus, this notation is seldom used in general mathematical or scientific settings.

Practically speaking, the notation $\mapsto$ ("maps to", an arrow with a bar at its tail) is flexible and convenient. It can be used to briefly mention and define a function without assigning it a name. In other cases, to define a function in full, a mapping defined with a "maps to" arrow could be stacked, in parallel, immediately below the declaration of the function name, domain, and the codomain. For example,

$$f : \mathbb{N} \to \mathbb{Z}$$
$$x \mapsto 4 - x.$$

The first part can be read as:

- "f is a function from $\mathbb{N}$ (the set of natural numbers) to $\mathbb{Z}$ (the set of integers)" or

- "f is a $\mathbb{Z}$-valued function of an $\mathbb{N}$-valued variable".

The second part is read:

- "x maps to $4 - x$."

In other words, this function has the natural numbers as domain, the integers as codomain. Strictly speaking, a function is properly defined only when the domain and codomain are specified. Moreover, the function

$$g : \mathbb{Z} \to \mathbb{Z}$$
$$x \mapsto 4 - x.$$

(with different domain) is a different function, even though the formulas defining f and g agree. Similarly, a function with a different codomain is also a different function. Nevertheless, many authors do not specify the domain and codomain, especially if these are clear from context. So in this example many just write $f(x) = 4 - x$. Sometimes, the maximal possible domain within a larger set implied by context is implicitly understood: a formula such as $f(x) = \sqrt{x^2 - 5x + 6}$ may mean that the domain of f is the set of real numbers x where the square root is defined (in this case $x \le 2$ or $x \ge 3$). However, in a different context, this expression might refer to a complex-valued function $f : \mathbb{R} \to \mathbb{C}$.

Finally, *dot notation* is occasionally used to specify a function by replacement of the variable of interest in an expression with a dot. For example, $a(\cdot)^2$ may stand for $x \mapsto ax^2$, and $\int_a^{(\cdot)} f(u)du$ may stand for the integral function $x \mapsto \int_a^x f(u)du$.

Specifying a Function

A function can be defined by any mathematical condition relating each argument (input value) to the corresponding output value. If the domain is finite, a function f may be defined by simply tabulating all the arguments x and their corresponding function values $f(x)$. More commonly, a function is defined by a formula, or (more generally) an algorithm — a recipe that tells how to compute the value of $f(x)$ given any x in the domain.

There are many other ways of defining functions. Examples include piecewise definitions, induction or recursion, algebraic or analytic closure, limits, analytic continuation, infinite series, and as solutions to integral and differential equations. The lambda calculus provides a powerful and flexible syntax for defining and combining functions of several variables. In advanced mathematics, some functions exist because of an axiom, such as the Axiom of Choice.

Graph

The *graph* of a function is its set of ordered pairs F. This is an abstraction of the idea of a graph as a picture showing the function plotted on a pair of coordinate axes; for example, $(3, 9)$, the point above 3 on the horizontal axis and to the right of 9 on the vertical axis, lies on the graph of $y = x^2$.

Formulas and Algorithms

Different formulas or algorithms may describe the same function. For instance $f(x) = (x + 1)(x - 1)$ is exactly the same function as $f(x) = x^2 - 1$. Furthermore, a function need not be described by a formula, expression, or algorithm, nor need it deal with numbers at all: the domain and codomain of a function may be arbitrary sets. One example of a function that acts on non-numeric inputs takes English words as inputs and returns the first letter of the input word as output.

As an example, the factorial function is defined on the nonnegative integers and produces a non-negative integer. It is defined by the following inductive algorithm: 0! is defined to be 1, and $n!$ is defined to be $n(n - 1)!$ for all positive integers n. The factorial function is denoted with the exclamation mark (serving as the symbol of the function) after the variable (postfix notation).

Computability

Functions that send integers to integers, or finite strings to finite strings, can sometimes be defined by an algorithm, which gives a precise description of a set of steps for computing the output of the function from its input. Functions definable by an algorithm are called *computable functions*. For example, the Euclidean algorithm gives a precise process to compute the greatest common divisor of two positive integers. Many of the functions studied in the context of number theory are computable.

Fundamental results of computability theory show that there are functions that can be precisely defined but are not computable. Moreover, in the sense of cardinality, almost all functions from the integers to integers are not computable. The number of computable functions from integers to integers is countable, because the number of possible algorithms is. The number of all functions from integers to integers is higher: the same as the cardinality of the real numbers. Thus most functions from integers to integers are not computable. Specific examples of uncomputable functions are known, including the busy beaver function and functions related to the halting problem and other undecidable problems.

Basic Properties

There are a number of general basic properties and notions. In this section, f is a function with domain X and codomain Y.

Image and Preimage

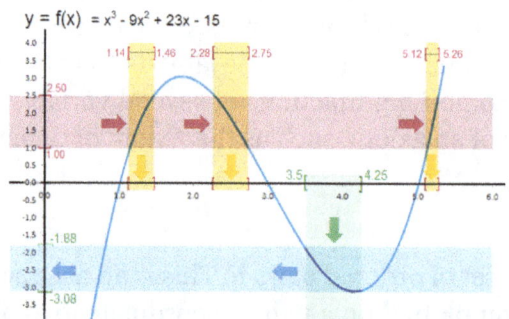

The graph of the function $f(x) = x^3 - 9x^2 + 23x - 15$. The interval $A = [3.5, 4.25]$ is a subset of the domain, thus it is shown as part of the x-axis (green). The image of A is (approximately) the interval $[-3.08, -1.88]$. It is obtained by projecting to the y-axis (along the blue arrows) the intersection of the graph with the light green area consisting of all points whose x-coordinate is between 3.5 and 4.25. the part of the (vertical) y-axis shown in blue. The preimage of $B = [1, 2.5]$ consists of three intervals. They are obtained by projecting the intersection of the light red area with the graph to the x-axis.

If A is any subset of the domain X, then $f(A)$ is the subset of the codomain Y consisting of all images of elements of A. We say the $f(A)$ is the *image* of A under f. The *image* of f is given by $f(X)$. On the other hand, the *inverse image* (or *preimage, complete inverse image*) of a subset B of the codomain Y under a function f is the subset of the domain X defined by

$$f^{-1}(B) = \{x \in X : f(x) \in B\}.$$

So, for example, the preimage of {4, 9} under the squaring function is the set {−3,−2,2,3}. The term range usually refers to the image, but sometimes it refers to the codomain.

By definition of a function, the image of an element x of the domain is always a single element y of the codomain. However, the preimage of a singleton set (a set with exactly one element) may in general contain any number of elements. For example, if $f(x) = 7$ (the constant function taking value 7), then the preimage of {5} is the empty set but the preimage of {7} is the entire domain. It is customary to write $f^{-1}(b)$ instead of $f^{-1}(\{b\})$, i.e.

$$f^{-1}(b) = \{x \in X : f(x) = b\}.$$

This set is sometimes called the fiber of b under f. (This notation is the same as that for the *inverse function*. However, the inverse function $f^{-1} : Y \to X$ of a function $f : X \to Y$ is defined if and only if the function is *one-to-one* and *onto*.

Use of $f(A)$ to denote the image of a subset $A \subseteq X$ is consistent so long as no subset of the domain is also an element of the domain. In some fields (e.g., in set theory, where ordinals are also sets of ordinals) it is convenient or even necessary to distinguish the two concepts; the customary notation is $f[A]$ for the set $\{f(x): x \in A\}$. Likewise, some authors use square brackets to avoid confusion between the inverse image and the inverse function. Thus they would write $f^{-1}[B]$ and $f^{-1}[b]$ for the preimage of a set and a singleton.

Injective and Surjective Functions

A function is called *injective* (or *one-to-one; 1-1*) if $f(a) \neq f(b)$ for any two elements $a, b, a \neq b$ of the domain. It is called *surjective* (or *onto*) if the range is identical to the codomain; that is, $f(X) = Y$. In other words, every element y in the codomain is mapped to by f from some x in the domain. Finally f is called *bijective* (or the function is a *one-to-one correspondence*) if it is both injective and surjective. A function that is injective, surjective, or bijective is referred to as an injection, a surjection, or a bijection, respectively. The existence of injections, surjections, or bijections between sets is the key concept defining the relative cardinalities (sizes) of the sets.

"One-to-one" and "onto" are terms that were more common in the older English language literature; "injective", "surjective", and "bijective" were originally coined as French words in the second quarter of the 20th century by the Bourbaki group and imported into English. As a word of caution, "a one-to-one function" is one that is injective, while a "one-to-one correspondence" refers to a bijective function. Also, the statement "f maps A onto B" differs from "f maps A into B" in that the former implies that f is an onto function (i.e., surjective), while the latter makes no assertion about the nature of the mapping. In more complicated statements the one letter difference can easily be missed. Due to the confusing nature of this older terminology, these terms have declined in popularity relative to the Bourbakian terms.

The above "color-of-the-shape" function is not injective, since two distinct shapes (the red triangle and the red rectangle) are assigned the same value. Moreover, it is not surjective, since the image of the function contains only three, but not all five colors in the codomain.

Function Composition

The *function composition* of two functions takes the output of one function as the input of a second one. More specifically, the composition of f with a function $g: Y \to Z$ is the function $g \circ f : X \to Z$ defined by

$$(g \circ f)(x) = g(f(x)).$$

That is, the value of x is obtained by first applying f to x to obtain $y = f(x)$ and then applying g to y to obtain $z = g(y)$. In the notation $g \circ f$, the function on the right, f, acts first and the function on the left, g acts second, reversing English reading order. The notation can be memorized by reading the notation as "g of f" or "g after f". The composition $g \circ f$ is only defined when the codomain of f is the domain of g. Assuming that, the composition in the opposite order $f \circ g$ need not be defined. Even if it is, i.e., if the codomain of f is the codomain of g, it is *not* in general true that

$$g \circ f = f \circ g.$$

That is, the order of the composition is important. For example, suppose $f(x) = x^2$ and $g(x) = x+1$. Then $g(f(x)) = x^2+1$, while $f(g(x)) = (x+1)^2$, which is x^2+2x+1, a different function.

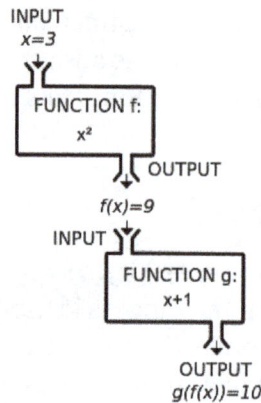

A composite function $g(f(x))$ can be visualized as the combination of two "machines". The first takes input x and outputs $f(x)$. The second takes as input the value $f(x)$ and outputs $g(f(x))$.

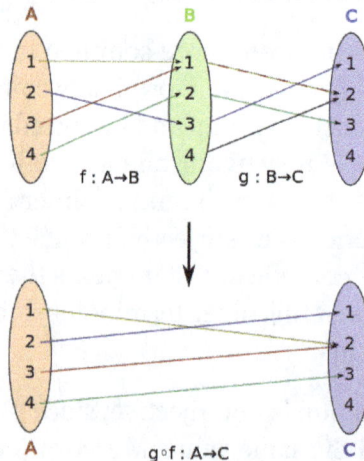

A concrete example of a function composition.

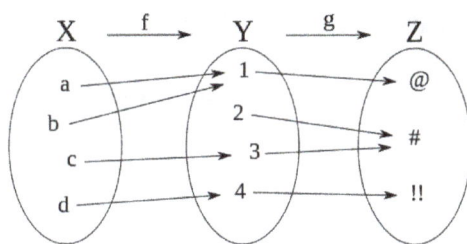

Another composition. For example, we have here $g \circ f \, (c) = \#$.

Identity Function

The unique function over a set X that maps each element to itself is called the *identity function* for X, and typically denoted by id_X. Each set has its own identity function, so the subscript cannot be omitted unless the set can be inferred from context. Under composition, an identity function is "neutral": if f is any function from X to Y, then

$$f \circ \mathrm{id}_X = f,$$
$$\mathrm{id}_Y \circ f = f.$$

Restrictions and Extensions

Informally, a *restriction* of a function f is the result of trimming its domain. More precisely, if S is any subset of X, the restriction of f to S is the function $f|_S$ from S to Y such that $f|_S(s) = f(s)$ for all s in S. If g is a restriction of f, then it is said that f is an *extension* of g.

The *overriding* of $f: X \to Y$ by $g: W \to Y$ (also called *overriding union*) is an extension of g denoted as $(f \oplus g): (X \cup W) \to Y$. Its graph is the set-theoretical union of the graphs of g and $f|_{X \setminus W}$. Thus, it relates any element of the domain of g to its image under g, and any other element of the domain of f to its image under f. Overriding is an associative operation; it has the empty function as an identity element. If $f|_{X \cap W}$ and $g|_{X \cap W}$ are pointwise equal (e.g., the domains of f and g are disjoint), then the union of f and g is defined and is equal to their overriding union. This definition agrees with the definition of union for binary relations.

Inverse Function

An *inverse function* for f, denoted by f^{-1}, is a function in the opposite direction, from Y to X, satisfying

$$f \circ f^{-1} = \mathrm{id}_Y, f^{-1} \circ f = \mathrm{id}_X.$$

That is, the two possible compositions of f and f^{-1} need to be the respective identity maps of X and Y.

As a simple example, if f converts a temperature in degrees Celsius C to degrees Fahrenheit F, the function converting degrees Fahrenheit to degrees Celsius would be a suitable f^{-1}.

$$f(C) = \frac{9}{5}C + 32$$
$$f^{-1}(F) = \frac{5}{9}(F - 32)$$

Such an inverse function exists if and only if f is bijective. In this case, f is called invertible. The notation $g \circ f$ (or, in some texts, just gf) and f^{-1} are akin to multiplication and reciprocal notation. With this analogy, identity functions are like the multiplicative identity, 1, and inverse functions are like reciprocals (hence the notation).

Types of Functions

Real-valued Functions

A real-valued function f is one whose codomain is the set of real numbers or a subset thereof. If, in addition, the domain is also a subset of the reals, f is a real valued function of a real variable. The study of such functions is called real analysis.

Real-valued functions enjoy so-called pointwise operations. That is, given two functions

$$f, g \colon X \to Y$$

where Y is a subset of the reals (and X is an arbitrary set), their (pointwise) sum $f+g$ and product $f \cdot g$ are functions with the same domain and codomain. They are defined by the formulas:

$$(f + g)(x) = f(x) + g(x),$$
$$(f \cdot g)(x) = f(x) \cdot g(x).$$

In a similar vein, complex analysis studies functions whose domain and codomain are both the set of complex numbers. In most situations, the domain and codomain are understood from context, and only the relationship between the input and output is given, but if $f(x) = \sqrt{x}$, then in real variables the domain is limited to non-negative numbers.

The following table contains a few particularly important types of real-valued functions:

Linear function	Quadratic function
A linear function.	A quadratic function.
$f(x) = ax + b.$	$f(x) = ax^2 + bx + c.$

Discontinuous function	Trigonometric functions
The signum function is not continuous, since it "jumps" at 0.	The sine and cosine functions.
Roughly speaking, a continuous function is one whose graph can be drawn without lifting the pen.	$f(x) = \sin(x)$ (red), $f(x) = \cos(x)$ (blue).

Multivariate Functions

A multivariate function is one which takes several inputs.

Further Types of Functions

There are many other special classes of functions that are important to particular branches of mathematics, or particular applications. Here is a partial list:

- differentiable, integrable
- polynomial, rational
- algebraic, transcendental
- odd or even
- convex, monotonic
- holomorphic, meromorphic, entire
- vector-valued
- computable

Function Spaces

The set of all functions from a set X to a set Y is denoted by $X \to Y$, by $[X \to Y]$, or by Y^X. The latter notation is motivated by the fact that, when X and Y are finite and of size $|X|$ and $|Y|$, then the number of functions $X \to Y$ is $|Y^X| = |Y|^{|X|}$. This is an example of the convention from enumerative combinatorics that provides notations for sets based on their cardinalities. If X is infinite and there is more than one element in Y then there are uncountably many functions from X to Y, though only countably many of them can be expressed with a formula or algorithm.

Currying

An alternative approach to handling functions with multiple arguments is to transform them into a chain of functions that each takes a single argument. For instance, one can interpret Add(3,5) to mean "first produce a function that adds 3 to its argument, and then apply the 'Add 3' function to 5". This transformation is called currying: Add 3 is curry(Add) applied to 3. There is a bijection between the function spaces $C^{A \times B}$ and $(C^B)^A$.

When working with curried functions it is customary to use prefix notation with function application considered left-associative, since juxtaposition of multiple arguments—as in $(f\,x\,y)$—naturally maps to evaluation of a curried function. Conversely, the $\rightarrow$ and $\mapsto$ symbols are considered to be right-associative, so that curried functions may be defined by a notation such as $f: Z \rightarrow Z \rightarrow Z = x \mapsto y \mapsto x{\cdot}y$.

Variants and Generalizations

Alternative Definition of a Function

The above definition of "a function from X to Y" is generally agreed on, however there are two different ways a "function" is normally defined where the domain X and codomain Y are not explicitly or implicitly specified. Usually this is not a problem as the domain and codomain normally will be known. With one definition saying the function defined by $f(x) = x^2$ on the reals does not completely specify a function as the codomain is not specified, and in the other it is a valid definition.

In the other definition a function is defined as a set of ordered pairs where each first element only occurs once. The domain is the set of all the first elements of a pair and there is no explicit codomain separate from the image. Concepts like surjective have to be refined for such functions, more specifically by saying that a (given) function is *surjective on a (given) set* if its image equals that set. For example, we might say a function f is surjective on the set of real numbers.

If a function is defined as a set of ordered pairs with no specific codomain, then $f: X \rightarrow Y$ indicates that f is a function whose domain is X and whose image is a subset of Y. This is the case in the ISO standard. Y may be referred to as the codomain but then any set including the image of f is a valid codomain of f. This is also referred to by saying that "f maps X into Y" In some usages X and Y may subset the ordered pairs, e.g. the function f on the real numbers such that $y=x^2$ when used as in $f: [0,4] \rightarrow [0,4]$ means the function defined only on the interval [0,2]. With the definition of a function as an ordered triple this would always be considered a partial function.

An alternative definition of the composite function $g(f(x))$ defines it for the set of all x in the domain of f such that $f(x)$ is in the domain of g. Thus the real square root of $-x^2$ is a function only defined at 0 where it has the value 0.

Functions are commonly defined as a type of relation. A relation from X to Y is a set of ordered pairs (x,y) with $x \in X$ and $y \in Y$. A function from X to Y can be described as a relation from X to Y that is left-total and right-unique. However, when X and Y are not specified there is a disagreement about the definition of a relation that parallels that for functions. Normally a relation is just defined as a set of ordered pairs and a correspondence is defined as a triple (X,Y,F), however the distinction between the two is often blurred or a relation is never referred to without specifying the two sets. The definition of a function as a triple defines a function as a type of correspondence, whereas the definition of a function as a set of ordered pairs defines a function as a type of relation.

Many operations in set theory, such as the power set, have the class of all sets as their domain, and therefore, although they are informally described as functions, they do not fit the set-theoretical definition outlined above, because a class is not necessarily a set. However some definitions of relations and functions define them as classes of pairs rather than sets of pairs and therefore do include the power set as a function.

Partial and Multi-valued Functions

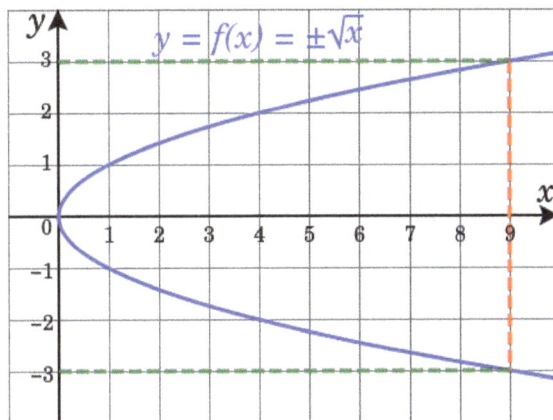

$f(x) = \pm\sqrt{x}$ is not a function in the proper sense, but a multi-valued function: it assigns to each positive real number x two values: the (positive) square root of x, and $-\sqrt{x}$.

In some parts of mathematics, including recursion theory and functional analysis, it is convenient to study *partial functions* in which some values of the domain have no association in the graph; i.e., single-valued relations. For example, the function f such that $f(x) = 1/x$ does not define a value for $x = 0$, since division by zero is not defined. Hence f is only a partial function from the real line to the real line. The term total function can be used to stress the fact that every element of the domain does appear as the first element of an ordered pair in the graph.

In other parts of mathematics, non-single-valued relations are similarly conflated with functions: these are called *multivalued functions*, with the corresponding term single-valued function for ordinary functions.

Functions with Multiple Inputs and Outputs

The concept of function can be extended to an object that takes a combination of two (or more) argument values to a single result. This intuitive concept is formalized by a function whose domain is the Cartesian product of two or more sets.

For example, consider the function that associates two integers to their product: $f(x, y) = x \cdot y$. This function can be defined formally as having domain Z×Z, the set of all integer pairs; codomain Z; and, for graph, the set of all pairs $((x, y), x \cdot y)$. Note that the first component of any such pair is itself a pair (of integers), while the second component is a single integer.

The function value of the pair (x, y) is $f((x, y))$. However, it is customary to drop one set of parentheses and consider $f(x, y)$ a function of two variables, x and y. Functions of two variables may be plotted on the three-dimensional Cartesian as ordered triples of the form $(x, y, f(x, y))$.

The concept can still further be extended by considering a function that also produces output that is expressed as several variables. For example, consider the integer divide function, with domain Z×N and codomain Z×N. The resultant (quotient, remainder) pair is a single value in the codomain.

Binary Operations

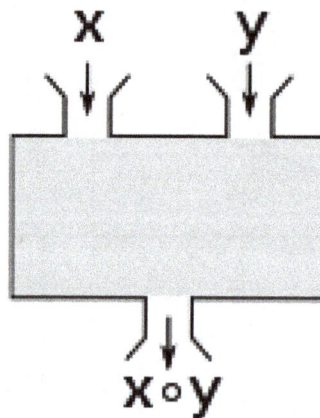

A binary operation is a function which assigns to each pair x and y its result $x \circ y$.

The familiar binary operations of arithmetic, addition and multiplication, can be viewed as functions from $\mathbb{R} \times \mathbb{R}$ to $\mathbb{R}$. This view is generalized in abstract algebra, where n-ary functions are used to model the operations of arbitrary algebraic structures. For example, an abstract group is defined as a set X and a function f from $X \times X$ to X that satisfies certain properties.

Traditionally, addition and multiplication are written in the infix notation: $x+y$ and $x \times y$ instead of $+(x, y)$ and $\times(x, y)$.

Functors

The idea of structure-preserving functions, or homomorphisms, led to the abstract notion of morphism, the key concept of category theory. In fact, functions $f : X \to Y$ are the morphisms in the category of sets, including the empty set: if the domain X is the empty set, then the subset of $X \times Y$ describing the function is necessarily empty, too. However, this is still a well-defined function. Such a function is called an empty function. In particular, the identity function of the empty set is defined, a requirement for sets to form a category.

The concept of categorification is an attempt to replace set-theoretic notions by category-theoretic ones. In particular, according to this idea, sets are replaced by categories, while functions between sets are replaced by functors.

Definition (Function):1. Let A and B be two sets. Then a function f : A→B is a rule that assigns to each element of A exactly one element of B.

2. The set A is called the domain of the function f.

3. The set B is called the co-domain of the function f.

The readers should carefully read the following important remark before proceeding further.

Remark: 1. If A = ø, then by convention, one assumes that there is a function, called the empty function, from A to B.

2. If B = ø then it can be easily observed that there is no function from A to B.

3. Some books use the word "map" in place of "function". So, both the words may be used interchangeably throughout the notes.

4. Throughout these notes, whenever the phrase "let f : A→B be a function" is used, it will be assumed that both A and B are non-empty sets.

Example: Let A = {a, b, c}, B = {1, 2, 3} and C = {3, 4}. Then verify that the examples given below are indeed functions.

    (a)   f : A→B, defined by f (a) = 3, f (b) = 3 and f (c) = 3.

    (b)   f : A→B, defined by f (a) = 3, f (b) = 2 and f (c) = 2.

    (c)   f : A→B, defined by f (a) = 3, f (b) = 1 and f (c) = 2.

    (d)   f : A→C, defined by f (a) = 3, f (b) = 3 and f (c) = 3.

    (e)   f : C→A, defined by f (3) = a, f (4) = c.

2. Verify that the following examples give functions, $f : \mathbb{Z} \to \mathbb{Z}$.

    (a) $f(x)=1$, if x is even and f (x) = 5, if x is odd.

    (b) $f(x)=-1$, for all $x \in \mathbb{Z}$.

    (c) $f(x)=x$ (mod 10), for all $\in \mathbb{Z}$.

    (d) $f(x)=1,$ if $x>0$, $f(0)=0$ and $f(x)=1, if\ x<0$.

Definition: Let f : A→B be a function. Then,

1. for each $x \in A$, the element $f\ (x) \in B$ is called the image of x under f.

2. the range/image of A under f equals f (A) = {f (a) : a ∈ A}.

3. the function f is said to be one-to-one if "for any two distinct elements $a_1, a_2 \in A$, $f(a_1) \neq f(a_2)$".

4. the function f is said to be onto if "for every element b ∈ B there exists an element a ∈ A, such that f (a) = b".

5. for any function g : B→C, the composition g ∘ f : A→C is a function defined by (g ∘ f)(a) = g(f (a)), for every a ∈ A.

Example: Let $f : \mathbb{N} \to \mathbb{Z}$ be defined by $f(x) = \begin{cases} \dfrac{-x}{2}, & \text{if } x \text{ is even,} \\ \dfrac{x+1}{2}, & \text{if } x \text{ is odd.} \end{cases}$ Then prove that f is one-one. Is f onto?

Solution: Let us use the contrapostive argument to prove that f is one-one. Let if possible $f(x) = f(y)$, for some $x, y \in \mathbb{N}$. Using the definition, one sees that x and y are either both odd or both even. So, let us assume that both x and y are even. In this case, $\dfrac{-x}{2} = \dfrac{-y}{2}$ and hence $x = y$. A similar argument holds, in case both x and y are odd.

Claim: f is onto.

Let $x \in \mathbb{Z}$ with $x \geq 1$. Then $2x - 1 \in \mathbb{N}$ and $f(2x - 1) = \dfrac{(2x-1)+1}{2} = x$. If $x \in \mathbb{N}$ and $x \leq 0$, then $-2x \in \mathbb{N}$ and $f(-2x) = \dfrac{-(-2x)}{2} = x$. Hence, f is indeed onto.

2. Let $f : \mathbb{N} \to \mathbb{Z}$ and $g : \mathbb{Z} \to \mathbb{Z}$ be defined by, $f(x) = 2x$ and $g(x) = \begin{cases} 0, & \text{if } x \text{ is odd,} \\ x/2, & \text{if } x \text{ is even,} \end{cases}$ respectively. Then prove that the functions f and $g \circ f$ are one-one but g is not one-one.

Solution: By definiton, it is clear that f is indeed one-one and g is not one-one. But

$$g \circ f(x) = g(f(x)) = g(2x) = \frac{(2x)}{2} = x,$$

for all $x \in \mathbb{N}$. Hence, $g \circ f : \mathbb{N} \to \mathbb{Z}$ is also one-one.

The next theorem gives some result related with composition of functions.

Theorem: (Properties of Functions). Consider the functions f : A→B, g : B→C and h : C→D.

1. Then $(h \circ g) \circ f = h \circ (g \circ f)$ (associativity holds).

2. If f and g are one-to-one then the function $g \circ f$ is also one-to-one.

3. If f and g are onto then the function $g \circ f$ is also onto.

Proof. First note that $g \circ f : A \to C$ and both $(h \circ g) \circ f$, $h \circ (g \circ f)$ are functions from A to D.

Proof of Part 1: The first part is direct, as for each a ∈ A,

$((h \circ g) \circ f)(a) = (h \circ g)(f(a)) = h(g(f(a))) = h((g \circ f)(a)) = (h \circ (g \circ f))(a)$.

Proof of Part 2: Need to show that "whenever $(g \circ f)(a_1) = (g \circ f)(a_2)$, for some $a_1, a_2 \in A$ then $a_1 = a_2$".

So, let us assume that $g(f(a_1)) = (g \circ f)(a_1) = (g \circ f)(a_2) = g(f(a_2))$, for some $a_1, a_2 \in A$. As g is one-one, the assumption gives $f(a_1) = f(a_2)$. But f is also one-one and hence $a_1 = a_2$.

Proof of Part 3: To show that "given any c ∈ C, there exists a ∈ A such that (g ∘f)(a) = c".

As g is onto, for the given c ∈ C, there exists b ∈ B such that g(b) = c. But f is also given to be onto. Hence, for the b obtained in previous step, there exists a ∈ A such that f (a) = b. Hence, we see that c = g(b) = g(f (a)) = (g ∘ f)(a).

Definition (Identity Function): Fix a set A and let e_A : A→A be defined by e_A(a) = a, for all a ∈ A. Then the function e_A is called the identity function or map on A.

The subscript A in Definition will be removed, whenever there is no chance of confusion about the domain of the function.

Theorem: (Properties of Identity Function) Fix two non-empty sets A and B and let f : A→B and g : B→A be any two functions. Also, let e : A→A be the identity map defined above. Then

1. e is a one-one and onto map.

2. the map f ∘ e = f .

3. the map e ∘ g = g.

Proof. Proof of Part 1: Since e(a) = a, for all a ∈ A, it is clear that e is one-one and onto.

Proof of Part 2: BY definition, (f ∘ e)(a) = f (e(a)) = f (a), for all a ∈ A. Hence, f ∘ e = f .

Proof of Part 3: The readers are advised to supply the proof.

Example: Let $f, g :$ ℕ → ℕ be defined by.

$$f(x) = 2x \text{ and } g(x) = \begin{cases} 0, & \text{if } x \text{ is odd}, \\ x/2, & \text{if } x \text{ is even}. \end{cases}$$

Then verify that $g \circ f :$ ℕ → ℕ is the identity map, whereas f ∘ g maps even numbers to itself and maps odd numbers to 0.

Definition (Invertible Function): A function f : A→B is said to be invertible if there exists a function g : B→A such that the map

1. g ∘ f : A→A is the identity map on A, and

2. f ∘ g : B→B is the identity map on B.

Let us now prove that if f : A→B is an invertible map then the map g : B→A, defined above is unique.

Theorem: Let f : A→B be an invertible map. Then the map

1. g defined in Definition is unique. The map g is generally denoted by f⁻¹.

2. (f⁻¹)⁻¹ = f .

Proof. The proof of the second part is left as an exercise for the readers. Let us now proceed with the proof of the first part.

Suppose g, h : B→A are two maps satisfying the conditions in Definition. Therefore, $g \circ f = e_A = h \circ f$ and $f \circ g = e_B = f \circ h$. Hence, using associativity of functions, for each b ∈ B, one has $g(b) = g(e_B(b)) = g((f \circ h)(b)) = (g \circ f)(h(b)) = e_A(h(b)) = h(b)$. Hence, the maps h and g are the same and thus the proof of the first part is over.

Theorem: Let f : A→B be a function. Then f is invertible if and only if f is one-one and onto.

Proof. Let f be invertible. To show, f is one-one and onto.

Since, f is invertible, there exists the map f^{-1} : B→A such that $f \circ f^{-1} = e_B$ and $f^{-1} \circ f = e_A$. So, now suppose that $f(a_1) = f(a_2)$, for some a_1, a_2 ∈ A. Then, using the map f^{-1}, we get

$$a_1 = e_A(a_1) = (f^{-1} \circ f)(a_1) = f^{-1}(f(a_1)) = f^{-1}(f(a_2)) = (f^{-1} \circ f)(a_2) = e_A(a_2) = a_2.$$

Thus, f is one-one. To prove onto, let b ∈ B. Then, by definition, $f^{-1}(b)$ ∈ A and $f(f^{-1}(b))$ =

$(f \circ f^{-1})(b) = e_B(b) = b$. Hence, f is onto as well.

Now, let us assume that f is one-one and onto. To show, f is invertible. Consider the map f^{-1}: B→A defined by "$f^{-1}(b) = a$ whenever f (a) = b", for each b ∈ B. This map is well-defined as f is onto and onto (note that onto implies that for each b ∈ B, there exists a ∈ A such that f (a) = b. Also, f is one-one implies that the element a obtained in the previous line is unique).

Now, it can be easily verified that $f \circ f^{-1} = e_B$ and $f^{-1} \circ f = e_A$ and hence f is indeed invertible.

We now state the following important theorem whose proof is beyond the scope of this book. The theorem is popularly known as the "Cantor-Bernstein-Schroeder theorem".

Definition (Cantor-Bernstein-Schroeder Theorem) Let A and B be two sets. If there exist injective (one-one) functions f : A→B (i.e., |A| ≤ |B|) and g : B→A (i.e., |A| ≥ |B|), then there exists a bijective (one-one and onto) function h : A→B (i.e., |A| = |B|).

Let A and B be two non-empty finite disjoint subsets of a set S. Then

1. |A ∪ B| = |A| + |B|.

2. |A × B| = |A| • |B|.

3. A and B have the same cardinality if there exists a one-one and onto function f : A→B.

Lemma: Let M and N be two sets such that |M| = m and |N| = n. Then the total number of functions f : M →N equals n^m.

Proof: Let M = $\{a_1, a_2, \ldots, a_m\}$ and N = $\{b_1, b_2, \ldots, b_n\}$. Since a function is determined as soon as we know the value of $f(a_i)$, for 1 ≤ i ≤ m, a function f : M →N has the form

$$f \leftrightarrow \begin{pmatrix} a_1 & a_2 & \cdots & a_m \\ f(a_1) & f(a_2) & \cdots & f(a_m) \end{pmatrix},$$

where $f(a_i)$ ∈ $\{b_1, b_2, \ldots, b_n\}$, for 1 ≤ i ≤ m. As there is no restriction on the function f, $f(a_i)$ has n choices, $b_1, b_2, \ldots, b_n$. Similarly, $f(a_2)$ has n choices, $b_1, b_2, \ldots, b_n$ and so on. Thus, the total number of functions $f : M \to N$ is

$$\underbrace{n \cdot n \cdots\cdots n}_{m \ tim \ es} = n^m.$$

Remark: Observe that Lemma is equivalent to the following question: In how many ways can m distinguishable/distinct balls be put into n distinguishable/distinct boxes?

Hint: Number the balls as $a_1, a_2, \ldots, a_m$ and the boxes as $b_1, b_2, \ldots, b_n$.

Lemma: Let M and N be two sets such that $|M| = m$ and $|N| = n$. Then the total number of distinct one-to-one functions $f : M \to N$ is $n(n-1)\ldots(n-m+1)$.

Proof: Observe that "f is one-to-one" means "whenever $x \neq y$ we must have $f(x) \neq f(y)$". Therefore, if m > n, then the number of such functions is 0.

So, let us assume that $m \leq n$ with $M = \{a_1, a_2, \ldots, a_m\}$ and $N = \{b_1, b_2, \ldots, b_n\}$. Then by definition, $f(a_1)$ has n choices, $b_1, b_2, \ldots, b_n$. Once $f(a_1)$ is chosen, there are only $n-1$ choices for $f(a_2)$ ($f(a_2)$ has to be chosen from the set $\{b_1, b_2, \ldots, b_n\} \setminus \{f(a_1)\}$). Similarly, there are only $n-2$ choices for $f(a_3)$ ($f(a_3)$ has to be chosen from the set $\{b_1, b_2, \ldots, b_n\} \setminus \{f(a_1), f(a_2)\}$), and so on. Thus, the required number is $n \cdot (n-1) \cdot (n-2) \ldots (n-m+1)$.

Remark: 1. The product $n(n-1) \ldots 3 \cdot 2 \cdot 1$ is denoted by n!, and is commonly called "n factorial".

2. By convention, we assume that $0! = 1$.

3. Using the factorial notation $n \cdot (n-1) \cdot (n-2) \cdots\cdots (n-m+1) = \dfrac{n!}{(n-m)!}$. This expression is generally denoted by $n_{(m)}$, and is called the falling factorial of n. Thus, if m > n then $n_{(m)} = 0$ and if n = m then $n_{(m)} = n!$.

4. The following conventions will be used in these notes:

$0! = 0_{(0)} = 1, 0^0 = 1, n_{(0)} = 1$ for all $n \geq 1$, $0_{(m)} = 0$ for $m \neq 0$.

The proof of the next corollary is immediate from Lemma and hence the proof is omitted.

Corollary: Let M and N be two sets such that $|M| = |N| = n$ (say). Then the number of one-to-one functions $f : M \to N$ equals n!, called "n-factorial".

Binomial Theorem

In elementary algebra, the binomial theorem (or binomial expansion) describes the algebraic expansion of powers of a binomial. According to the theorem, it is possible to expand the polynomial $(x + y)^n$ into a sum involving terms of the form $ax^b y^c$, where the exponents b and c are nonnegative integers with $b + c = n$, and the coefficient a of each term is a specific positive integer depending on n and b. For example,

$$(x+y)^4 = x^4 + 4x^3y + 6x^2y^2 + 4xy^3 + y^4.$$

$$1$$
$$1 \quad 1$$
$$1 \quad 2 \quad 1$$
$$1 \quad 3 \quad 3 \quad 1$$
$$1 \quad 4 \quad 6 \quad 4 \quad 1$$
$$1 \quad 5 \quad 10 \quad 10 \quad 5 \quad 1$$

The binomial coefficients appear as the entries of Pascal's triangle where each entry is the sum of the two above it.

The coefficient a in the term of ax^by^c is known as the binomial coefficient $\binom{n}{b}$ or $\binom{n}{c}$ (the two have the same value). These coefficients for varying n and b can be arranged to form Pascal's triangle. These numbers also arise in combinatorics, where $\binom{n}{b}$ gives the number of different combinations of b elements that can be chosen from an n-element set.

Special cases of the binomial theorem were known from ancient times. The 4th century B.C. Greek mathematician Euclid mentioned the special case of the binomial theorem for exponent 2. There is evidence that the binomial theorem for cubes was known by the 6th century in India.

Binomial coefficients, as combinatorial quantities expressing the number of ways of selecting k objects out of n without replacement, were of interest to the ancient Hindus. The earliest known reference to this combinatorial problem is the *Chandaśāstra* by the Hindu lyricist Pingala (c. 200 B.C.), which contains a method for its solution. The commentator Halayudha from the 10th century A.D. explains this method using what is now known as Pascal's triangle. By the 6th century A.D., the Hindu mathematicians probably knew how to express this as a quotient $\dfrac{n!}{(n-k)!k!}$, and a clear statement of this rule can be found in the 12th century text *Lilavati* by Bhaskara.

The binomial theorem as such can be found in the work of 11th-century Persian mathematician Al-Karaji, who described the triangular pattern of the binomial coefficients. He also provided a mathematical proof of both the binomial theorem and Pascal's triangle, using a primitive form of mathematical induction. The Persian poet and mathematician Omar Khayyam was probably familiar with the formula to higher orders, although many of his mathematical works are lost. The binomial expansions of small degrees were known in the 13th century mathematical works of Yang Hui and also Chu Shih-Chieh. Yang Hui attributes the method to a much earlier 11th century text of Jia Xian, although those writings are now also lost.

In 1544, Michael Stifel introduced the term "binomial coefficient" and showed how to use them to express $(1+a)^n$ in terms of $(1+a)^{n-1}$, via "Pascal's triangle". Blaise Pascal studied the eponymous triangle comprehensively in the treatise *Traité du triangle arithmétique* (1653). However, the pattern of numbers was already known to the European mathematicians of the late Renaissance, including Stifel, Niccolò Fontana Tartaglia, and Simon Stevin.

Isaac Newton is generally credited with the generalized binomial theorem, valid for any rational exponent.

Theorem Statement

According to the theorem, it is possible to expand any power of $x + y$ into a sum of the form

$$(x+y)^n = \binom{n}{0}x^n y^0 + \binom{n}{1}x^{n-1}y^1 + \binom{n}{2}x^{n-2}y^2 + \cdots + \binom{n}{n-1}x^1 y^{n-1} + \binom{n}{n}x^0 y^n,$$

where each $\binom{n}{k}$ is a specific positive integer known as a binomial coefficient. (When an exponent is zero, the corresponding power expression is taken to be 1 and this multiplicative factor is often omitted from the term. Hence one often sees the right side written as $\binom{n}{0}x^n + \ldots$.) This formula is also referred to as the binomial formula or the binomial identity. Using summation notation, it can be written as

$$(x+y)^n = \sum_{k=0}^{n}\binom{n}{k}x^{n-k}y^k = \sum_{k=0}^{n}\binom{n}{k}x^k y^{n-k}.$$

The final expression follows from the previous one by the symmetry of x and y in the first expression, and by comparison it follows that the sequence of binomial coefficients in the formula is symmetrical. A simple variant of the binomial formula is obtained by substituting 1 for y, so that it involves only a single variable. In this form, the formula reads

$$(1+x)^n = \binom{n}{0}x^0 + \binom{n}{1}x^1 + \binom{n}{2}x^2 + \cdots + \binom{n}{n-1}x^{n-1} + \binom{n}{n}x^n,$$

or equivalently

$$(1+x)^n = \sum_{k=0}^{n}\binom{n}{k}x^k.$$

Examples

```
                  1
               1     1
            1     2     1
         1     3     3     1
      1     4     6     4     1
   1     5    10    10     5     1
1     6    15    20    15     6     1
1  7   21   35   35   21   7   1
```
Pascal's triangle.

The most basic example of the binomial theorem is the formula for the square of $x + y$:

$$(x+y)^2 = x^2 + 2xy + y^2.$$

The binomial coefficients 1, 2, 1 appearing in this expansion correspond to the second row of Pascal's triangle. (Note that the top "1" of the triangle is considered to be row 0, by convention.) The coefficients of higher powers of $x + y$ correspond to lower rows of the triangle:

$$(x+y)^3 = x^3 + 3x^2y + 3xy^2 + y^3,$$
$$(x+y)^4 = x^4 + 4x^3y + 6x^2y^2 + 4xy^3 + y^4,$$
$$(x+y)^5 = x^5 + 5x^4y + 10x^3y^2 + 10x^2y^3 + 5xy^4 + y^5,$$
$$(x+y)^6 = x^6 + 6x^5y + 15x^4y^2 + 20x^3y^3 + 15x^2y^4 + 6xy^5 + y^6,$$
$$(x+y)^7 = x^7 + 7x^6y + 21x^5y^2 + 35x^4y^3 + 35x^3y^4 + 21x^2y^5 + 7xy^6 + y^7.$$

Several patterns can be observed from these examples. In general, for the expansion $(x + y)^n$:

1. the powers of x start at n and decrease by 1 in each term until they reach 0 (with $x^0 = 1$, often unwritten);

2. the powers of y start at 0 and increase by 1 until they reach n;

3. the nth row of Pascal's Triangle will be the coefficients of the expanded binomial when the terms are arranged in this way;

4. the number of terms in the expansion before like terms are combined is the sum of the coefficients and is equal to 2^n; and

5. there will be $n + 1$ terms in the expression after combining like terms in the expansion.

The binomial theorem can be applied to the powers of any binomial. For example,

$$(x+2)^3 = x^3 + 3x^2(2) + 3x(2)^2 + 2^3$$
$$= x^3 + 6x^2 + 12x + 8.$$

For a binomial involving subtraction, the theorem can be applied by using the form $(x - y)^n = (x + (-y))^n$. This has the effect of changing the sign of every other term in the expansion:

$$(x-y)^3 = (x+(-y))^3 = x^3 + 3x^2(-y) + 3x(-y)^2 + (-y)^3 = x^3 - 3x^2y + 3xy^2 - y^3.$$

Geometric Explanation

For positive values of a and b, the binomial theorem with $n = 2$ is the geometrically evident fact that a square of side $a + b$ can be cut into a square of side a, a square of side b, and two rectangles with sides a and b. With $n = 3$, the theorem states that a cube of side $a + b$ can be cut into a cube of side a, a cube of side b, three $a \times a \times b$ rectangular boxes, and three $a \times b \times b$ rectangular boxes.

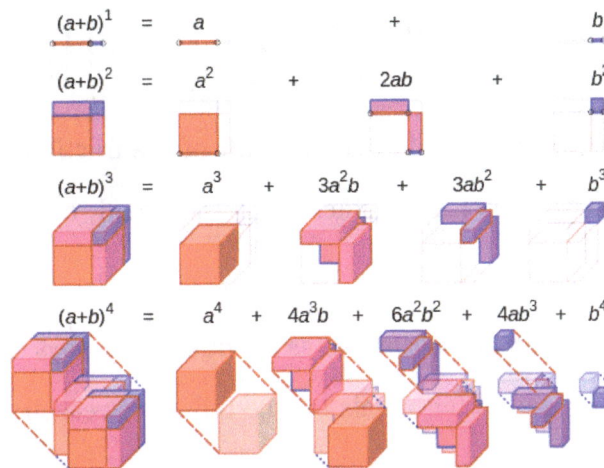

Visualisation of binomial expansion up to the 4th power.

In calculus, this picture also gives a geometric proof of the derivative $(x^n)' = nx^{n-1}$: if one sets $a = x$ and $b = \Delta x$, interpreting b as an infinitesimal change in a, then this picture shows the infinitesimal change in the volume of an n-dimensional hypercube, $(x + \Delta x)^n$, where the coefficient of the linear term (in Δx) is nx^{n-1}, the area of the n faces, each of dimension $(n-1)$:

$$(x + \Delta x)^n = x^n + nx^{n-1}\Delta x + \binom{n}{2}x^{n-2}(\Delta x)^2 + \cdots.$$

Substituting this into the definition of the derivative via a difference quotient and taking limits means that the higher order terms, ( ) and higher, become negligible, and yields the formula $(x^n)' = nx^{n-1}$, interpreted as

"the infinitesimal rate of change in volume of an n-cube as side length varies is the area of n of its $(n-1)$:-dimensional faces".

If one integrates this picture, which corresponds to applying the fundamental theorem of calculus, one obtains Cavalieri's quadrature formula, the integral $\int x^{n-1} dx = \dfrac{1}{n}x^n$.

Binomial Coefficients

The coefficients that appear in the binomial expansion are called binomial coefficients. These are usually written $\binom{n}{k}$, and pronounced "n choose k".

Formulae

The coefficient of $x^{n-k}y^k$ is given by the formula

$$\binom{n}{k} = \frac{n!}{k!(n-k)!}$$

which is defined in terms of the factorial function $n!$. Equivalently, this formula can be written

$$\binom{n}{k} = \frac{n(n-1)\cdots(n-k+1)}{k(k-1)\cdots 1} = \prod_{\ell=1}^{k} \frac{n-\ell+1}{\ell} = \prod_{\ell=0}^{k-1} \frac{n-\ell}{k-\ell}$$

with k factors in both the numerator and denominator of the fraction. Note that, although this formula involves a fraction, the binomial coefficient $\binom{n}{k}$ is actually an integer.

Combinatorial Interpretation

The binomial coefficient $\binom{n}{k}$ can be interpreted as the number of ways to choose k elements from an n-element set. This is related to binomials for the following reason: if we write $(x + y)^n$ as a product

$$(x+y)(x+y)(x+y)\cdots(x+y),$$

then, according to the distributive law, there will be one term in the expansion for each choice of either x or y from each of the binomials of the product. For example, there will only be one term x^n, corresponding to choosing x from each binomial. However, there will be several terms of the form $x^{n-2}y^2$, one for each way of choosing exactly two binomials to contribute a y. Therefore, after combining like terms, the coefficient of $x^{n-2}y^2$ will be equal to the number of ways to choose exactly 2 elements from an n-element set.

Proofs

Combinatorial Proof

Example

The coefficient of xy^2 in

$$\begin{aligned}
(x+y)^3 &= (x+y)(x+y)(x+y) \\
&= xxx + xxy + xyx + \underline{xyy} + yxx + \underline{yxy} + \underline{yyx} + yyy \\
&= x^3 + 3x^2y + \underline{3xy^2} + y^3.
\end{aligned}$$

equals $\binom{3}{2} = 3$ because there are three x,y strings of length 3 with exactly two y's, namely,

$$xyy, \ yxy, \ yyx,$$

corresponding to the three 2-element subsets of $\{1, 2, 3\}$, namely,

$$\{2,3\}, \{1,3\}, \{1,2\},$$

where each subset specifies the positions of the y in a corresponding string.

General Case

Expanding $(x + y)^n$ yields the sum of the 2^n products of the form $e_1 e_2 \ldots e_n$ where each e_i is x or y.

Rearranging factors shows that each product equals $x^{n-k}y^k$ for some k between 0 and n. For a given k, the following are proved equal in succession:

- the number of copies of $x^{n-k}y^k$ in the expansion

- the number of n-character x,y strings having y in exactly k positions

- the number of k-element subsets of $\{1, 2, ..., n\}$

- $\dbinom{n}{k}$ (this is either by definition, or by a short combinatorial argument if one is defining $\dbinom{n}{k}$ as $\dfrac{n!}{k!(n-k)!}$).

This proves the binomial theorem.

Inductive Proof

Induction yields another proof of the binomial theorem. When $n = 0$, both sides equal 1, since $x^0 = 1$ and $\dbinom{0}{0} = 1$. Now suppose that the equality holds for a given n; we will prove it for $n + 1$. For j, $k \geq 0$, let $[f(x, y)]_{j,k}$ denote the coefficient of $x^j y^k$ in the polynomial $f(x, y)$. By the inductive hypothesis, $(x + y)^n$ is a polynomial in x and y such that $[(x + y)^n]_{j,k}$ is $\dbinom{n}{k}$ if $j + k = n$, and 0 otherwise. The identity

$$(x+y)^{n+1} = x(x+y)^n + y(x+y)^n$$

shows that $(x + y)^{n+1}$ also is a polynomial in x and y, and

$$[(x+y)^{n+1}]_{j,k} = [(x+y)^n]_{j-1,k} + [(x+y)^n]_{j,k-1},$$

since if $j + k = n + 1$, then $(j - 1) + k = n$ and $j + (k - 1) = n$. Now, the right hand side is

$$\binom{n}{k} + \binom{n}{k-1} = \binom{n+1}{k},$$

by Pascal's identity. On the other hand, if $j + k \neq n + 1$, then $(j - 1) + k \neq n$ and $j + (k - 1) \neq n$, so we get $0 + 0 = 0$. Thus

$$(x+y)^{n+1} = \sum_{k=0}^{n+1} \binom{n+1}{k} x^{n+1-k} y^k,$$

which is the inductive hypothesis with $n + 1$ substituted for n and so completes the inductive step.

Generalizations

Newton's Generalized Binomial Theorem

Around 1665, Isaac Newton generalized the binomial theorem to allow real exponents other than nonnegative integers. (The same generalization also applies to complex exponents.) In this generalization, the finite sum is replaced by an infinite series. In order to do this, one needs to give meaning to binomial coefficients with an arbitrary upper index, which cannot be done using the usual formula with factorials. However, for an arbitrary number r, one can define

$$\binom{r}{k} = \frac{r(r-1)\cdots(r-k+1)}{k!} = \frac{(r)_k}{k!},$$

where $(\cdot)_k$ is the Pochhammer symbol, here standing for a falling factorial. This agrees with the usual definitions when r is a nonnegative integer. Then, if x and y are real numbers with $|x| > |y|$, and r is any complex number, one has

$$(x+y)^r = \sum_{k=0}^{\infty} \binom{r}{k} x^{r-k} y^k$$

$$= x^r + r x^{r-1} y + \frac{r(r-1)}{2!} x^{r-2} y^2 + \frac{r(r-1)(r-2)}{3!} x^{r-3} y^3 + \cdots.$$

When r is a nonnegative integer, the binomial coefficients for $k > r$ are zero, so this equation reduces to the usual binomial theorem, and there are at most $r + 1$ nonzero terms. For other values of r, the series typically has infinitely many nonzero terms.

For example, with $r = 1/2$ gives the following series for the square root:

$$\sqrt{1+x} = 1 + \frac{1}{2}x - \frac{1}{8}x^2 + \frac{1}{16}x^3 - \frac{5}{128}x^4 + \frac{7}{256}x^5 - \cdots$$

Taking $r = -1$, the generalized binomial series gives the geometric series formula, valid for $|x| < 1$:

$$(1+x)^{-1} = \frac{1}{1+x} = 1 - x + x^2 - x^3 + x^4 - x^5 + \cdots$$

More generally, with $r = -s$:

$$\frac{1}{(1-x)^s} = \sum_{k=0}^{\infty} \binom{s+k-1}{k} x^k \equiv \sum_{k=0}^{\infty} \binom{s+k-1}{s-1} x^k.$$

So, for instance, when $s = 1/2$,

$$\frac{1}{\sqrt{1+x}} = 1 - \frac{1}{2}x + \frac{3}{8}x^2 - \frac{5}{16}x^3 + \frac{35}{128}x^4 - \frac{63}{256}x^5 + \cdots$$

Further Generalizations

The generalized binomial theorem can be extended to the case where x and y are complex numbers. For this version, one should again assume $|x| > |y|$ and define the powers of $x + y$ and x using a holomorphic branch of log defined on an open disk of radius $|x|$ centered at x.

The generalized binomial theorem is valid also for elements x and y of a Banach algebra as long as $xy = yx$, x is invertible, and $||y/x|| < 1$.

Multinomial Theorem

The binomial theorem can be generalized to include powers of sums with more than two terms. The general version is

$$(x_1 + x_2 + \cdots + x_m)^n = \sum_{k_1 + k_2 + \cdots + k_m = n} \binom{n}{k_1, k_2, \ldots, k_m} x_1^{k_1} x_2^{k_2} \cdots x_m^{k_m}.$$

where the summation is taken over all sequences of nonnegative integer indices k_1 through k_m such that the sum of all k_i is n. (For each term in the expansion, the exponents must add up to n). The coefficients $\binom{n}{k_1, \cdots, k_m}$ are known as multinomial coefficients, and can be computed by the formula

$$\binom{n}{k_1, k_2, \ldots, k_m} = \frac{n!}{k_1! \cdot k_2! \cdots k_m!}.$$

Combinatorially, the multinomial coefficient $\binom{n}{k_1, \cdots, k_m}$ counts the number of different ways to partition an n-element set into disjoint subsets of sizes k_1, ..., k_m.

Multi-binomial Theorem

It is often useful when working in more dimensions, to deal with products of binomial expressions. By the binomial theorem this is equal to

$$(x_1 + y_1)^{n_1} \cdots (x_d + y_d)^{n_d} = \sum_{k_1=0}^{n_1} \cdots \sum_{k_d=0}^{n_d} \binom{n_1}{k_1} x_1^{k_1} y_1^{n_1-k_1} \cdots \binom{n_d}{k_d} x_d^{k_d} y_d^{n_d-k_d}.$$

This may be written more concisely, by multi-index notation, as

$$(x + y)^\alpha = \sum_{v \le \alpha} \binom{\alpha}{v} x^v y^{\alpha-v}.$$

Applications

Multiple-angle Identities

For the complex numbers the binomial theorem can be combined with De Moivre's formula to yield multiple-angle formulas for the sine and cosine. According to De Moivre's formula,

$$\cos(nx) + i\sin(nx) = (\cos x + i\sin x)^n.$$

Using the binomial theorem, the expression on the right can be expanded, and then the real and imaginary parts can be taken to yield formulas for $\cos(nx)$ and $\sin(nx)$. For example, since

$$\left(\cos x + i \sin x\right)^2 = \cos^2 x + 2i \cos x \sin x - \sin^2 x,$$

De Moivre's formula tells us that

$$\cos(2x) = \cos^2 x - \sin^2 x \quad \text{and} \quad \sin(2x) = 2\cos x \sin x,$$

which are the usual double-angle identities. Similarly, since

$$\left(\cos x + i \sin x\right)^3 = \cos^3 x + 3i \cos^2 x \sin x - 3\cos x \sin^2 x - i \sin^3 x,$$

De Moivre's formula yields

$$\cos(3x) = \cos^3 x - 3\cos x \sin^2 x \quad \text{and} \quad \sin(3x) = 3\cos^2 x \sin x - \sin^3 x.$$

In general,

$$\cos(nx) = \sum_{k \text{ even}} (-1)^{k/2} \binom{n}{k} \cos^{n-k} x \sin^k x$$

and

$$\sin(nx) = \sum_{k \text{ odd}} (-1)^{(k-1)/2} \binom{n}{k} \cos^{n-k} x \sin^k x.$$

Series for e

The number e is often defined by the formula

$$e = \lim_{n \to \infty} \left(1 + \frac{1}{n}\right)^n.$$

Applying the binomial theorem to this expression yields the usual infinite series for e. In particular:

$$\left(1 + \frac{1}{n}\right)^n = 1 + \binom{n}{1}\frac{1}{n} + \binom{n}{2}\frac{1}{n^2} + \binom{n}{3}\frac{1}{n^3} + \cdots + \binom{n}{n}\frac{1}{n^n}.$$

The kth term of this sum is

$$\binom{n}{k}\frac{1}{n^k} = \frac{1}{k!} \cdot \frac{n(n-1)(n-2)\cdots(n-k+1)}{n^k}.$$

As $n \to \infty$, the rational expression on the right approaches one, and therefore

$$\lim_{n\to\infty}\binom{n}{k}\frac{1}{n^k}=\frac{1}{k!}.$$

This indicates that e can be written as a series:

$$e=\sum_{k=0}^{\infty}\frac{1}{k!}=\frac{1}{0!}+\frac{1}{1!}+\frac{1}{2!}+\frac{1}{3!}+\cdots.$$

Indeed, since each term of the binomial expansion is an increasing function of n, it follows from the monotone convergence theorem for series that the sum of this infinite series is equal to e.

Derivative of the Power Function

In finding the derivative of the power function $f(x) = x^n$ for integer n using the definition of derivative, one can expand the binomial $(x + h)^n$.

Nth Derivative of a Product

To indicate the formula for the derivative of order n of the product of two functions, the formula of the binomial theorem is used symbolically.

Probability

The binomial theorem is closely related to the probability mass function of the negative binomial distribution. The probability of a (countable) collection of independent Bernoulli trials $\{X_t\}_{t\in S}$ with probability of success $p\in[0,1]$ all not happening is

$$P\left(\bigcap_{t\in S}X_t^C\right)=(1-p)^{|S|}=\sum_{n=0}^{|S|}\binom{|S|}{n}(-p)^n$$

A useful upper bound for this quantity is e^{-pn}.

The Binomial Theorem in Abstract Algebra

Formula (1) is valid more generally for any elements x and y of a semiring satisfying $xy = yx$. The theorem is true even more generally: alternativity suffices in place of associativity.

The binomial theorem can be stated by saying that the polynomial sequence $\{ 1, x, x^2, x^3, \dots \}$ is of binomial type.

- The binomial theorem is mentioned in the Major-General's Song in the comic opera The Pirates of Penzance.

- Professor Moriarty is described by Sherlock Holmes as having written a treatise on the binomial theorem.

- The Portuguese poet Fernando Pessoa, using the heteronym Álvaro de Campos, wrote that

"Newton's Binomial is as beautiful as the Venus de Milo. The truth is that few people notice it."

Lemma: Let N be a finite set consisting of n elements. Then the number of distinct subsets of N, of size k, $1 \le k \le n$, equals $\dfrac{n!}{(n-k)! \cdot k!}$

Proof: It can be easily verified that the result holds for k = 1. Hence, we fix a positive integer k, with $2 \le k \le n$. Then observe that any one-to-one function $f : \{1, 2, \ldots, k\} \to N$ gives rise to the following:

1. a set K = Im(f) = {f (i) : $1 \le i \le k$}. The set K is a subset of N and |K| = k (as f is one-to-one). Also,

2. given the set K = Im(f) = {f (i) : $1 \le i \le k$}, one gets a one-to-one function g : $\{1, 2, \ldots, k\} \to K$, defined by g(i) = f (i), for $1 \le i \le k$.

Therefore, we define two sets A and B by

A = {f : $\{1, 2, \ldots, k\} \to N$ | f is one-to-one}, and

B = {K ⊂ N | |K| = k} × {f : $\{1, 2, \ldots, k\} \to K$ | f is one-to-one}.

Thus, the above argument implies that there is a bijection between the sets A and B and therefore, it follows that |A| = |B|. Also, using Lemma, we know that |A| = $n_{(k)}$ and |B| = |{K ⊂ N | |K| = k}| × k!. Hence

Number of subsets of N of size $k = |\{K \subset N \mid |K| = k\}| = \dfrac{n_{(k)}}{k!} = \dfrac{n!}{(n-k)!.k!}$.

Remark: Let N be a set consisting of n elements.

1. Then, for n ≥ k, the number $\dfrac{n!}{k!(n-k)!}$ is generally denoted by $\dbinom{n}{k}$, and is called "n choose k". Thus, $\dbinom{n}{k}$ is a positive integer and equals "Number of subsets, of a set consisting of n elements, of size k".

2. Let K be a subset of N of size k. Then N \ K is again a subset of N of size n − k. Thus, there is one-to-one correspondence between subsets of size k and subsets of size n − k.

Thus, $\dbinom{n}{k} = \dbinom{n}{n-k}$.

3. The following conventions will be used:

$$\binom{n}{k} = \begin{cases} 0, & \text{if } n < k, \\ 1, & \text{if } k = 0. \end{cases}$$

Lemma: Fix a positive integer n. Then, for any two commuting symbols x and y

$$(x + y)^n = \sum_{k=0}^{n} \binom{n}{k} x^k y^{n-k}.$$

Proof: The expression $(x+y)^n = \underbrace{(x+y)\cdot(x+y)\cdots\cdots(x+y)}_{n\,times}$. Note that the above mul -tiplication is same as adding all the 2^n products (appearing due to the choice of either choosing x or choosing y, from each of the above n-terms). Since either x or y is chosen from each of the n-terms, the product looks like $x^k y^{n-k}$, for some choice of k, $o \le k \le n$. Therefore, for a fixed k, $o \le k \le n$, the term $x^k y^{n-k}$ appears $\binom{n}{k}$ times as we need to choose k places from n places, for x (and thus leaving n – k places for y), giving the expression $\binom{n}{k}$ as a coefficient of $x^k y^{n-k}$.

Hence, the required result follows.

Remark: Fix a positive integer n.

1. Then the numbers $\binom{n}{k}$ are called Binomial Coefficients as they appear in the expansion of $(x+y)^n$.

2. Substituting $x = y = 1$, one gets $2^n = \sum_{k=0}^{n} \binom{n}{k}$.

3. Observe that $(x+y+z)^n = \underbrace{(x+y+z)\cdot(x+y+z)\cdots\cdots(x+y+z)}_{n\,tims}$. Note that in this expression, we need to choose, say

(a) i places from the n possible places for x $(i \ge 0)$,

(b) j places from the remaining n – i places for $y (j \ge 0)$ and

thus leaving the n – i – j places for z (with n – i – j $\ge$ o). Hence, one has

$$(x+y+z)^n = \sum_{i,j\ge 0, i+j\le n} \binom{n}{i}\binom{n-i}{j} x^i y^j z^{n-i-j}.$$

4. The expression $\binom{n}{i}\binom{n-i}{j} = \dfrac{n!}{i!\,j!;(n-i-j)!}$ is also denoted by n $\binom{n}{i,j,n-i-j}$.

5. Similarly, if $i_1, i_2, \ldots, i_k$ are non-negative integers, such that $i_1 + i_2 + \ldots + i_k = n$, then the coefficient of $x_1^{i_1} x_2^{i_2} \cdots x_k^{i_k}$ in the expansion of $(x_1 + x_2 + \cdots + x_k)n$ equals

$$\binom{n}{i_1, i_2, \ldots, i_k} = \dfrac{n!}{i_1!\, i_2!\cdots i_k!}.$$

That is,

$$(x_1 + x_2 + \cdots + x_k)^n = \sum_{\substack{i_1, \ldots, i_k \ge 0 \\ i_1 + i_2 + \cdots + i_k = n}} \binom{n}{i_1, i_2, \ldots, i_k} x_1^{i_1} x_2^{i_2} \cdots x_k^{i_k}.$$

These coefficient and called multinomial coefficients.

Onto Functions and the Stirling Numbers of Second Kind

Definition: Let $|A| = n$. Then the number of partitions of the set A into m-parts is denoted by S(n, m). The symbol S(n, m) is called the Stirling number of the second kind.

The following conventions will be used:

$$S(n,m) = \begin{cases} 1, & \textit{if } n = m \\ 0, & \textit{if } n > 0, m = 0 \\ 0, & \textit{if } n < m. \end{cases}$$

2. If $n > m$ then a recursive method to compute the numbers S(n, m). A formula for the numbers S(n, m) is also given in equation above.

3. Consider the problem of determining the number of ways of putting m distinguishable/distinct balls into n indistinguishable boxes with the restriction that no box is empty.

Let $M = \{a_1, a_2, \ldots, a_m\}$ be the set of m distinct balls. Then, we observe the following:

(a) Since the boxes are indistinguishable, we can assume that the number of balls in each of the boxes is in a non-increasing order.

(b) Let A_i, for $1 \le i \le n$, denote the set of balls in the i-th box. Then $|A_1| \ge |A_2| \ge \ldots \ge |A_n|$ and

$$\bigcup_{i=1}^{n} A_i = M$$

(c) As each box is non-empty, each A_i is non-empty, for $1 \le i \le n$.

Thus, we see that we have obtained a partition of the set M, consisting of m elements, into n-parts, $A_1, A_2, \ldots, A_n$. Hence, the number of required ways is given by S(m, n), the Stirling number of second kind.

We are now ready to look at the problem of counting the number of onto functions $f : M \to N$. But to make the argument clear, we take an example.

Example: Let $f : \{a, b, c, d, e\} \to \{1, 2, 3\}$ be an onto function given by

$f(a) = f(b) = f(c) = 1, f(d) = 2$ and $f(e) = 3$.

Then this onto function, gives a partition $B_1 = \{a, b, c\}$, $B_2 = \{d\}$ and $B_3 = \{e\}$ of the set $\{a, b, c, d, e\}$ into 3-parts. Also, suppose that we are given a partition $A_1 = \{a, d\}$, $A_2 = \{b, e\}$ and $A_3 = \{c\}$ of $\{a, b, c, d, e\}$ into 3-parts. Then, this partition gives rise to the following 3! onto functions from $\{a, b, c, d, e\}$ into $\{1, 2, 3\}$:

$f_1(a) = f_1(d) = 1, f_1(b) = f_1(e) = 2, f_1(c) = 3,$ i.e., $f_1(A_1) = 1, f_1(A_2) = 2, f_1(A_3) = 3$

$f_2(a) = f_2(d) = 1, f_2(b) = f_2(e) = 3, f_2(c) = 2,$ i.e., $f_2(A_1) = 1, f_2(A_2) = 3, f_2(A_3) = 2$

$f_3(a) = f_3(d) = 2, f_3(b) = f_3(e) = 1, f_3(c) = 3,$ i.e., $f_3(A_1) = 2, f_3(A_2) = 1, f_3(A_3) = 3$

$f_4(a) = f_4(d) = 2, f_4(b) = f_4(e) = 3, f_4(c) = 1,$ i.e., $f_4(A_1) = 2, f_4(A_2) = 3, f_4(A_3) = 1$

$f_5(a) = f_5(d) = 3, f_5(b) = f_5(e) = 1, f_5(c) = 2,$ i.e., $f_5(A_1) = 3, f_5(A_2) = 1, f_5(A_3) = 2$

$f_6(a) = f_6(d) = 3, f_6(b) = f_6(e) = 2, f_6(c) = 1,$ i.e., $f_6(A_1) = 3, f_6(A_2) = 2, f_6(A_3) = 1.$

Lemma: Let M and N be two finite sets with $|M| = m$ and $|N| = n$. Then the total number of onto functions $f : M \to N$ is $n!S(m, n)$.

Proof: By definition, "f is onto" implies that "for all $y \in N$ there exists $x \in M$ such that $f(x) = y$. Therefore, the number of onto functions $f : M \to N$ is 0, whenever m < n. So, let us assume that m $\geq$ n and $N = \{b_1, b_2, \ldots, b_n\}$. Then, we observe the following:

1. Fix i, $1 \leq i \leq n$. Then $f^{-1}(b_i) = \{x \in M \mid f(x) = b_i\}$ is a non-empty set as f is an onto function.

2. $f^{-1}(b_i) \cap f^{-1}(b_j) = \emptyset$, whenever $1 \leq i \neq j \leq n$ as f is a function.

3. $\bigcup_{i=1}^{n} f^{-1}(b_i) = M$ as the domain of f is M .

Therefore, if we write $A_i = f^{-1}(b_i)$, for $1 \leq i \leq n$, then $A_1, A_2, \ldots, A_n$ gives a partition of M into n-parts. Also, for $1 \leq i \leq n$ and $x \in A_i$, we note that $f(x) = b_i$. That is, for $1 \leq i \leq n$, $|f(A_i)| = |\{b_i\}| = 1$.

Conversely, each onto function $f : M \to N$ is completely determined by

- a partition, say $A_1, A_2, \ldots, A_n$, of M into $n = |N|$ parts, and

- a one-to-one function $g : \{A_1, A_2, \ldots, A_n\} \to N$, where $f(x) = b_i$, whenever $x \in A_j$ and $g(A_j) = b_i$.

Hence,

$$\left|\{f : M \to N : f \text{ is onto}\}\right| = \left|\{g : \{A_1, A_2, \ldots, A_n\} \to N : g \text{ is one-to-one}\}\right|$$
$$\times \left|\text{Partition of } M \text{ into } n - \text{parts}\right|$$
$$= n!S(m, n). \tag{1}$$

Lemma: Let m and n be two positive integers and let $\ell = \min\{m, n\}$. Then

$$n^m = \sum_{k=1}^{\ell} \binom{n}{k} k! S(m, k). \tag{2}$$

Proof: Let M and N be two sets with $|M| = m$ and $|N| = n$ and let A denote the set of all functions $f : M \to N$. We compute $|A|$ using two different methods to get Equation (2).

The first method uses Lemma to give $|A| = n^m$. The second method uses the idea of onto functions. Let $f_0 : M \to N$ be any function and let $K = f_0(M) = \{f_0(x) : x \in M\} \subset N$.

Then, using f_0, we define a function $g : M \to K$, by $g(x) = f_0(x)$, for all $x \in M$. Then clearly g is an onto function with $|K| = k$ for some k, $1 \le k \le \ell = \min\{m, n\}$. Thus, $A = \bigcup_{k=1}^{\ell} A_k$, where $A_k = \{f : M \to N \mid |f(M)| = k\}$, for $1 \le k \le \ell$. Note that $A_k \cap A_j = \emptyset$, whenever

$1 \le j \ne k \le \ell$. Now, using Lemma, a subset of N of size k can be selected in $\binom{n}{k}$ ways. Thus, for $1 \le k \le \ell$

$$|A_k| = \left|\{K : K \subset N, \ |K| = k\}\right| \times \left|\{f : M \to K \mid f \text{ is onto}\}\right| = \binom{n}{k} k! S(m, k).$$

Therefore,

$$|A| = \left|\bigcup_{k=1}^{\ell} A_i\right| = \sum_{k=1}^{\ell} |A_k| = \sum_{k=1}^{\ell} \binom{n}{k} k! S(m, k).$$

Remark: The numbers S(m, k) can be recursively calculated using Equation (2).

(a) For example, taking $n \ge 1$ and substituting $m = 1$ in Equation (2) gives

$$n = n^1 = \sum_{k=1}^{\ell} \binom{n}{k} k! S(1, k) = n. \ 1!. S(1,1)$$

.

Thus, $S(1, 1) = 1$. Now, using n = 1 and $m \ge 2$ in Equation (2) gives

$$1 = 1^m = \sum_{k=1}^{\ell} \binom{1}{k} k! S(m, k) = 1. \ 1!. \ S(m,1).$$

Hence, the above two calculations implies that S(m, 1) = 1 for all $m \ge 1$.

(b) Use this to verify that S(5, 2) = 15, S(5, 3) = 25, ; S(5, 4) = 10, S(5, 5) = 1.

2. The problem of counting the total number of onto functions $f : M \to N$, with $|M| = m$ and $|N| = n$ is similar to the problem of determining the number of ways to put m distinguishable/distinct balls into n distinguishable/distinct boxes with the restriction that no box is empty.

Example: Determine the number of ways to seat 4 couples in a row if each couple seats together.

Solution: A couple can be thought of as one cohesive group (they are to be seated together). So, the 4 cohesive groups can be arranged in 4! ways. But a couple can sit either as "wife and husband" or "husband and wife". So, the total number of arrangements is 2^4 4!.

Indistinguishable Balls and Distinguishable Boxes

Example: Consider the following three problems which have the same solution.

1. Determine the number of distinct strings that can be formed using 3 A's and 6 B's.

2. Determine the number of solutions to the equation $x_1 + x_2 + x_3 + x_4 = 6$, where each $x_i \in \mathbb{Z}$ and $0 \le x_i \le 6$.

3. Determine the number of ways of placing 6 indistinguishable balls into 4 distinguishable boxes.

Solution: The solution is based on the understanding that all the three problems correspond to forming strings using +'s (or |'s) and 1's (or balls) in place of A'a and B's.

$$BBABBBABA \qquad 11+111+1+ = 2+3+1+0 \qquad \bullet\bullet|\bullet\bullet\bullet|\bullet$$

$$ABBBBBAAB \qquad +11111++1 = 0+5+0+1 \qquad |\bullet\bullet\bullet\bullet\bullet||\bullet|$$

$$ABBBABABB \qquad +111+1+11 = 0+3+1+2 \qquad |\bullet\bullet\bullet|\bullet|\bullet\bullet$$

Understanding the three problems

Note that the 3 A's are indistinguishable among themselves and the same holds for 6 B's. Thus, we need to find 3 places, from the 9 = 3 + 6 places, for the A's. Hence, the answer is $\binom{9}{3}$. The answer will remain the same as we just need to replace A's with +'s (or |'s) and B's with 1's (or balls) in any string of 3 A's and 6 B's. Figure above or note that four numbers can be added using 3 +'s or four adjacent boxes can be created by putting 3 vertical lines or |'s.

We now generalize this example to a general case.

Lemma: Determine the number of

1. solutions to the equation $x_1 + x_2 + \cdots + x_n = m$, where each $x_i \in \mathbb{Z}$ and $0 \le x_i \le m$.

2. ways to put m indistinguishable balls into n distinguishable boxes.

Proof: Note that the number m or the m balls can be replaced with m 1's or m ⋆'s. Once this is done, using the idea in example, we see that it is enough to find the number of distinct strings formed using n − 1 +'s (or |'s) and m 1's (or m ⋆'s). Then the indistinguishability of the +'s (or |'s) and placing them among the 1's (or ⋆'s), we get

$$\binom{n-1+m}{m} = \binom{n-1+m}{n-1} = \binom{m+(n-1)}{m}.$$

Remark: Observe that the problems in Lemma is same as " Determine the number of non-decreasing sequences of length m using the numbers 1, 2, . . . , n". Hint: Since we are looking at a non-decreasing sequences, we note that the sequence is determined if we know the number of times a particular number has appeared in the sequence. So, let x_i, for $1 \le i \le n$, denote the number of times the number i has appeared in the sequence.

Graph of a Function

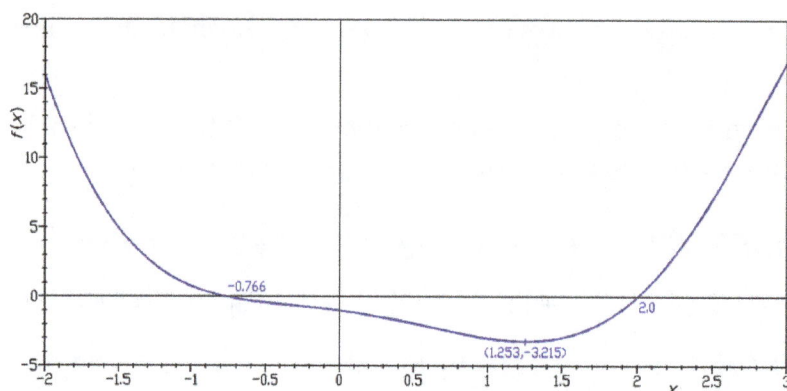

Graph of the function f(x) = x⁴ − 4ˣ over the interval [−2,+3]. Also shown are its two real roots and global minimum over the same interval.

In mathematics, the graph of a function f is the collection of all ordered pairs (x, f(x)). If the function input x is a scalar, the graph is a two-dimensional graph, and for a continuous function is a curve. If the function input x is an ordered pair (x_1, x_2) of real numbers, the graph is the collection of all ordered triples $(x_1, x_2, f(x_1, x_2))$, and for a continuous function is a surface.

Informally, if x is a real number and f is a real function, graph may mean the graphical representation of this collection, in the form of a line chart: a curve on a Cartesian plane, together with Cartesian axes, etc. Graphing on a Cartesian plane is sometimes referred to as curve sketching. The graph of a function on real numbers may be mapped directly to the graphic representation of the function. For general functions, a graphic representation cannot necessarily be found and the formal definition of the graph of a function suits the need of mathematical statements, e.g., the closed graph theorem in functional analysis.

The concept of the graph of a function is generalized to the graph of a relation. Note that although a function is always identified with its graph, they are not the same because it will happen that two functions with different codomain could have the same graph. For example, the cubic polynomial mentioned below is a surjection if its codomain is the real numbers but it is not if its codomain is the complex field.

To test whether a graph of a curve is a function of x, one uses the vertical line test. To test whether a graph of a curve is a function of y, one uses the horizontal line test. If the function has an inverse, the graph of the inverse can be found by reflecting the graph of the original function over the line y = x.

In science, engineering, technology, finance, and other areas, graphs are tools used for many purposes. In the simplest case one variable is plotted as a function of another, typically using rectangular axes.

In the modern foundation of mathematics known as set theory, a function and its graph are essentially the same thing.

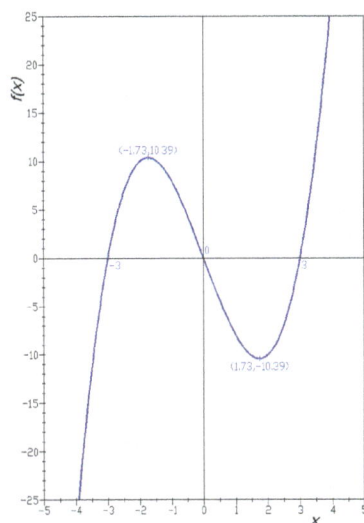

Graph of the function f(x) = x³ – 9

Examples

Functions of one Variable

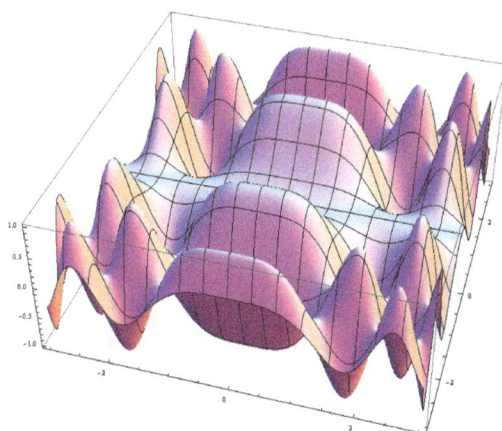

Graph of the function f(x, y) = sin(x²) · cos(y²).

The graph of the function.

$$f(x) = \begin{cases} a, & \text{if } x = 1, \\ d, & \text{if } x = 2, \\ c, & \text{if } x = 3, \end{cases}$$

is

$$\{(1, a), (2, d), (3, c)\}.$$

The graph of the cubic polynomial on the real line

$$f(x) = x^3 - 9x$$

is

$$\{(x, x^3 - 9x) : x \text{ is a real number}\}.$$

If this set is plotted on a Cartesian plane, the result is a curve.

Functions of Two Variables

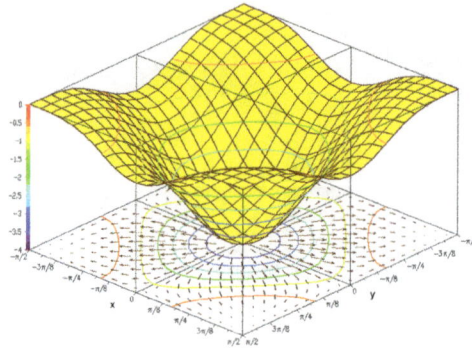

Plot of the graph of f(x, y) = −(cos(x²) + cos(y²))², also showing its gradient projected on the bottom plane.

The graph of the trigonometric function

$$f(x, y) = \sin(x^2)\cos(y^2)$$

is

$$\{(x, y, \sin(x^2)\cos(y^2)) : x \text{ and } y \text{ are real numbers}\}.$$

If this set is plotted on a three dimensional Cartesian coordinate system, the result is a surface.

Oftentimes it is helpful to show with the graph, the gradient of the function and several level curves. The level curves can be mapped on the function surface or can be projected on the bottom plane. The second figure shows such a drawing of the graph of the function:

$$f(x, y) = -(\cos(x^2) + \cos(y^2))^2$$

Normal to a Graph

Given a function f of n variables: $x_1, \ldots, x_n$, the normal to the graph is

$$(\nabla f, -1)$$

(up to multiplication by a constant). This is seen by considering the graph as a level set of the function $g(x, z) = f(x) - z$, and using that ∇g is normal to the level sets.

Generalizations

The graph of a function is contained in a Cartesian product of sets. An X–Y plane is a cartesian product of two lines, called X and Y, while a cylinder is a cartesian product of a line and a circle, whose height, radius, and angle assign precise locations of the points. Fibre bundles aren't cartesian products, but appear to be up close. There is a corresponding notion of a graph on a fibre bundle called a section.

Tools for Plotting Function Graphs

Hardware

- Graphing calculator

- Oscilloscope

Lattice Paths and Catalan Numbers

Consider a lattice of integer lines in the plane. The set $S = \{(m, n) : m, n = 0, 1, 2, \ldots\}$ are said to be the points of the lattice and the lines joining these points are called the edges of the lattice. Now, let us fix two points in this lattice, say (m_1, n_1) and m_2, n_2, with $m_2 \geq m_1$ and $n_2 \geq n_1$. Then we define an increasing/lattice path from (m_1, n_1) to (m_2, n_2) to be a subset $\{e_1, e_2, \ldots, e_k\}$ of S such that

1. either $e_1 = (m_1, n_1 + 1)$ or $e_1 = (m_1 + 1, n_1)$;

2. either $e_k = (m_2, n_2 - 1)$ or $e_k = (m_2 - 1, n_2)$; and

3. if we represent the tuple $e_i = (a_i, b_i)$, for $1 \leq i \leq k$, then for $2 \leq j \leq k$,

    (a) either $a_j = a_{j-1}$ and $b_j = b_{j-1} + 1$

    (b) or $b_j = b_{j-1}$ and $a_j = a_{j-1} + 1$.

That is, the movement on the lattice is either to the RIGHT or UP (see figure below). Now, let us look at some of the questions related to this topic.

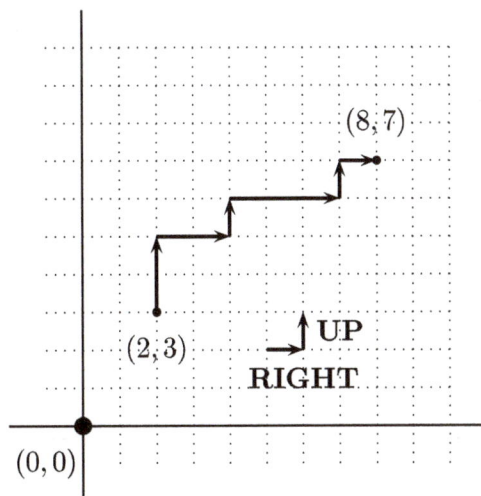

A lattice with a lattice path from (2, 3) to (8, 7).

Determine the number of lattice paths from (0, 0) to (m, n).

Solution: Note that at each stage, the coordinates of the lattice path increases, by exactly one positive value, either in the X-coordinate or in the Y-coordinate. Therefore, to reach (m, n) from (0, 0), the total increase in the X-direction is exactly m and in the Y-direction is exactly n. That is, each lattice path is a sequence of length m + n, consisting of m R's (movement along X-axis/RIGHT) and n U's (movement along the Y-axis/UP). So, we need to find m places for the R's among the m + n places (R and U together). Thus, the required answer is $\binom{m+n}{m}$.

2. Use the method of lattice paths to prove the following result on Binomial Coefficients:

$$\sum_{\ell=0}^{m} \binom{n+l}{l} = \binom{n+m+1}{m}.$$

Solution: Observe that the right hand side corresponds to the number of lattice paths from (0, 0) to (m, n + 1), whereas the left hand side corresponds to the number of lattice paths from (0, 0) to (ℓ, n), where $0 \leq \ell \leq m$.

Now, fix ℓ, $0 \leq \ell \leq m$. Then to each lattice path from (0, 0) to (ℓ, n), say P, we adjoin the path $Q = \underbrace{U\ RR \cdots R}_{m-\ell\ times}$. Then the path P $\cup$ Q, corresponding to a lattice path from (0, 0) to (ℓ, n) and from (ℓ, n) to (ℓ, n + 1) and finally from (ℓ, n + 1) to (m, n + 1), gives a lattice path from (0, 0) to (m, n + 1). These lattice paths, as we vary ℓ, for $0 \leq \ell \leq m$, are all distinct and hence the result follows.

Determine the number of lattice paths from (0, 0) to (n, n) that do not go above the line Y = X (see figure below).

Solution: The first move from (0, 0) is R (corresponding to moving to the point (1, 0)) as we are not allowed to go above the line Y = X. So, in principle, all our lattice paths are from (1, 0) to (n,n) with the condition that these paths do not cross the line Y = X.

Using Example, the total number of lattice paths from (1, 0) to (n, n) is $\binom{2n-1}{n}$. So, we need to subtract from $\binom{2n-1}{n}$ a number, say N_0, where N_0 equals the number of lattice paths from (1, 0) to (n, n) that cross the line Y = X.

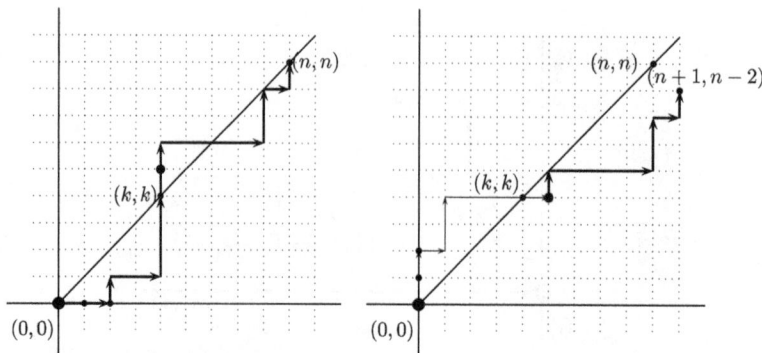

Lattice paths giving the mirror symmetry from (0,0) to (k, k).

To compute the value of N_0, we decompose each lattice path that crosses the line $Y = X$ into two sub-paths. Let P be a path from $(1, 0)$ to (n, n) that crosses the line $Y = X$. Then this path crosses the line $Y = X$ for the first time at some point, say (k, k), $1 \leq k \leq n - 1$.

Claim: there exists a 1-1 correspondence between lattice paths from $(1, 0)$ to (n, n) that crosses the line $Y = X$ and the lattice paths from $(0, 1)$ to $(n + 1, n - 1)$.

Let $P = P_1 P_2 \ldots P_{2k-1} P_{2k} P_{2k+1} \ldots P_{2n-1}$ be the path from $(1, 0)$ to (n, n) that crosses the line $Y = X$. Then $P_1, P_2, \ldots, P_{2k-2}$ consist of a sequence of $(k - 1)$ R's and $(k - 1)$ U's. Also, $P_{2k-1} = P_{2k} = U$ and the sub-path $P_{2k+1} P_{2k+2} \ldots P_{2n-1}$ consist of a sequence that has $(n - k)$ R's and $(n - k - 1)$ U's. Also, for any i, $1 \leq i \leq 2k - 2$, in the sub-path $P_1 P_2 \ldots P_i$, # of R's $= |\{j : P_j = R, 1 \leq j \leq i\}| \geq |\{\ell : P_\ell = U, 1 \leq \ell \leq i\}| = $ # of U's.   (1.1)

Now, P is mapped to a path Q, such that $Q_i = \begin{cases} P_i, & \text{if } 2k+1 \leq i \leq 2n-1, \\ \{R,U\} \setminus P_i, & \text{if } 1 \leq i \leq 2k. \end{cases}$

Then we see that the path Q consists of exactly $(k - 1) + 2 + (n - k) = n + 1$ R's and $(k - 1) + (n - k - 1) = n - 2$ U's. Also, the condition that $Q_i = \{R, U\} \setminus P_i$, for $1 \leq i \leq 2k$, implies that the path Q starts from the point $(0, 1)$. Therefore, Q is a path that starts from $(0, 1)$ and consists of $(n + 1)$ R's and $(n - 2)$ U's and hence Q ends at the point $(n + 1, n - 1)$.

Also, if Q' is a path from $(0, 1)$ to $(n + 1, n - 1)$, then Q' consists of $(n + 1)$ R's and $(n - 2)$ U's. So, in any such sequence an instant occurs when the number of R's exceeds the number of U's by 2. Suppose this occurrence happens for the first time at the $(2k)^{\text{th}}$ instant, for some k, $1 \leq k \leq n - 1$. Then there are $(k + 1)$ R's and $(k - 1)$ U's till the first $(2k)^{\text{th}}$ instant and $(n - k)$ R's and $(n - 1 - k)$ U's in the remaining part of the sequence. So, Q' can be replaced by a path P', such that

$$(P')_i = \begin{cases} (Q')_i, & \text{if } 2k+1 \leq i \leq 2n-1, \\ \{R,U\} \setminus (Q')_i, & \text{if } 1 \leq i \leq 2k. \end{cases}$$

It can be easily verified that P' is a lattice path from $(1, 0)$ to (n, n) that crosses the line $Y = X$. Thus, the proof of the claim is complete.

Hence, the number of lattice paths from $(1, 0)$ to (n, n) that crosses the line $Y = X$ equals the number of lattice paths from $(0, 1)$ to $(n + 1, n - 1)$. But, usinge example, the number of lattice paths from $(0, 1)$ to $(n + 1, n - 1)$ equals $\binom{2n-1}{n+1}$. Hence, the number of lattice paths from $(0, 0)$ to (n, n) that does not go above the line $Y + X$ is

$$\binom{2n-1}{n} - \binom{2n-1}{n+1} = \frac{1}{n+1}\binom{2n}{n}.$$

This number is popularly known as the nth Catalan Number, denoted C_n.

Catalan Numbers

The book titled "Enumerative Combinatorics" by Stanley gives a comprehensive list of places in combinatorics where Catalan numbers appear. A few of them are mentioned here for the inquisitive mind.

1. Suppose in an election two candidates A and B get exactly n votes. Then C_n equals the number of ways of counting the votes such that candidate A is always ahead of candidate B.

For example,

$C_3 = 5 = |\{AAABBB, AABABB, AABBAB, ABAABB, ABABAB\}|$.

2. Suppose, we need to multiply n + 1 given numbers, say $a_1, a_2, \ldots, a_{n+1}$. Then the different ways of multiplying these numbers, without changing the order of the elements, equals C_n.

For example,

$C_3 = 5 = |\{(((a_1 a_2)a_3)a_4), ((a_1 a_2)(a_3 a_4)), ((a_1(a_2 a_3))a_4), (a_1((a_2 a_3)a_4)), (a_1(a_2(a_3 a_4)))\}|$.

3. C_n also equals the number of ways that a convex (n+2)-gon cab be sub-divided into triangles by its diagonals so that no two diagonals intersect inside the (n + 2)-gon. For example, for a pentagon, the different ways are

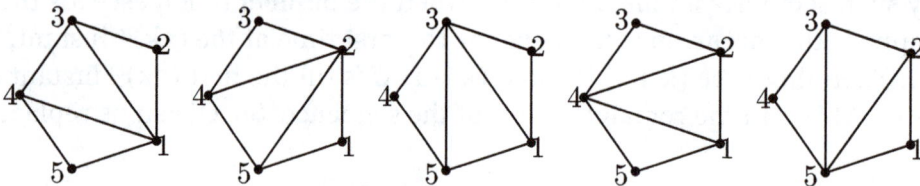

Different divisions of pentagon.

4. C_n is also equal to the number of full binary trees on 2n + 1 vertices, where recall that a full binary tree is a rooted binary tree in which every node either has exactly two offsprings or has no offspring.

Full binary trees on 7 vertices (or 4 leaves).

5. C_n is also equal to the number of Dyck paths from (0, 0) to (2n, 0), where recall that a Dyck path is a movement on an integer lattice in which each step is either in the North East or in the South East direction (so the only movement from (0, 0) is either to (1, 1) or to (1, −1)).

6. C_n also equals the number of n nonintersecting chords joining 2n points on the circumfer- ence of a circle.

Non-intersecting chords using 6 points on the circle.

7. C_n also equals the number of integer sequences that satisfy $1 \leq a_1 \leq a_2 \leq \cdots \leq a_n$ and $a_i \leq i$, for all i, $1 \leq i \leq n$.

Let A_n denote the set of all lattice paths from $(0, 0)$ to (n, n) and let $B_n \subset A_n$ denote the set of all lattice paths from $(0, 0)$ to (n, n) that does not go above the line $Y = X$. Then, the numerical values of $|A_n|$ and $|B_n|$ imply that $(n + 1) \cdot |B_n| = |A_n|$. The question arises: can we find a partition of the set A_n into $(n + 1)$-parts, say $S_0, S_1, S_2, \ldots, S_n$ such that $S_0 = B_n$ and $|S_i| = |S_0|$, for $1 \leq i \leq n$?

The answer is in affirmative. Observe that any path from $(0, 0)$ to (n, n) has n right moves. So, the path is specified as soon as we know the successive right moves $R_1, R_2, \ldots, R_n$, where R_i equals ℓ if and only if R_i lies on the line $Y = \ell$. For example, in the third figure, $R_1 = 0$, $R_2 = 0$, $R_3 = 1$, $R_4 = 1$, ... These R_i's satisfy

$$0 \leq R_1 \leq R_2 \leq \cdots \leq R_n \leq n.$$

That is, any element of A_n can be represented by an ordered n-tuple $(R_1, R_2, \ldots, R_n)$ satisfying the above equation. Conversely, it can be easily verified that any ordered n-tuple $(R_1, R_2, \ldots, R_n)$ satisfying the equation corresponds to a lattice path from $(0, 0)$ to (n, n). Note that, using Remark, among the above n-tuples, the tuples that satisfy $R_i \leq i - 1$, for $1 \leq i \leq n$ are elements of B_n, and vice-versa. Now, we use the tuples that represent the elements of B_n to get $n + 1$ maps, $f_0, f_1, \ldots, f_n$, in such a way that $f_j(B_n)$ and $f_k(B_n)$ are disjoint, for $0 \leq j \neq k \leq n$ and $A_n = \bigcup_{k=0}^{n} f_k(B_n)$. In particular, for a fixed k, $0 \leq k \leq n$, the map $f_k : B_n \to A_n$ is defined by

$$f_k\left((R_1, R_2, \ldots, R_n)\right) = \left(R_{i_1} \oplus_{n+1} k, R_{i_2} \oplus_{n+1} k, \ldots, R_{i_n} \oplus_{n+1} k\right),$$

where $\oplus_{n+1}$ denotes addition modulo $n + 1$ and $i_1, i_2, \ldots, i_n$ is a rearrangement of the numbers 1, 2, . . . , n such that $0 \leq R_{i_1} \oplus_{n+1} k \leq R_{i_2} \oplus_{n+1} k \leq \cdots \leq R_{i_n} \oplus_{n+1} k$. The readers are advised to prove the following exercise as they give the required partition of the set A_n.

Generalizations

1. Let n, k be non-negative integers with $0 \leq k \leq n$. Then in Lemma, "the number of ways of choosing a subset of size k from a set consisting of n elements" was denoted by the binomial coefficients, $\binom{n}{k} = \dfrac{n!}{k!(n-k)!}$. Since, for each k, $0 \leq k \leq n$, $(n - k)!$ divides $n!$, let us think of

$\binom{n}{k} = \dfrac{n.(n-1)\cdots(n-k+1)}{k!}$. With this understanding, the numbers $\binom{n}{k}$ can be generalized. That

is, in the generalized form, for any $n \in \mathbb{C}$ and for any non-negative integer k, one has

$$\binom{n}{k} = \begin{cases} 0, & \text{if } k < 0 \\ 0, & \text{if } n = 0, \; n \neq k \\ 1, & \text{if } n = k \\ \dfrac{n \cdot (n-1) \cdots (n-k+1)}{k!}, & \text{otherwise.} \end{cases} \qquad (3)$$

With the notations as above, one has the following theorem which is popularly known as the generalized binomial theorem. We state it without proof.

Theorem: Let n be any real number. Then

$$(1+x)^n = 1 + \binom{n}{1}x + \binom{n}{2}x^2 + \cdots + \binom{n}{r}x^r + \cdots.$$

In particular, $(1-x)^{-1} = 1 + x + x^2 + x^3 + \cdots$ and if $a, \; b \in \mathbb{R}$ with $|a| < |b|$, then

$$(a+b)^n = b^n\left(1+\frac{a}{b}\right)^n = b^n\sum_{r \geq 0}\binom{n}{r}\left(\frac{a}{b}\right)^r = \sum_{r \geq 0}\binom{n}{r}a^r b^{n-r}.$$

Let us now understand Theorem through the following examples.

(a) Let $n = \dfrac{1}{2}$. In this case, for $k \geq 1$, Equation (3) gives

$$\binom{\frac{1}{2}}{k} = \frac{\frac{1}{2}\cdot\left(\frac{1}{2}-1\right)\cdots\left(\frac{1}{2}-k+1\right)}{k!} = \frac{1\cdot(-1)\cdot(-3)\cdots(3-2k)}{2^k k!} = \frac{(-1)^{k-1}(2k-2)!}{2^{2k-1}(k-1)!k!}.$$

Thus,

$$(1+x)^{1/2} = \sum_{k \geq 0}\binom{\frac{1}{2}}{k}x^k = 1 + \frac{1}{2}x + \frac{-1}{2^3}x^2 + \frac{-1}{2^4}x^3 + \sum_{k \geq 4}\frac{(-1)^{k-1}(2k-2)!}{2^{2k-1}(k-1)!k!}x^k.$$

This can also be obtained using the Taylor series expansion of $f(x) = (1+x)^{1/2}$ around $x = 0$. Recall that the Taylor series expansion of $f(x)$ around $x = 0$ equals

$f(x) = f(0) + f'(0)x + \dfrac{f''(0)}{2!}x^2 + \sum_{k \geq 3}\dfrac{f(k)(0)}{k!}x^k$, where $f(0) = 1$, $f'(0) = \dfrac{1}{2}$, $f''(0) = \dfrac{-1}{2^2}$ and

in general $f^{(k)}(0) = \dfrac{1}{2}\cdot\left(\dfrac{1}{2}-1\right)\cdots\left(\dfrac{1}{2}-k+1\right)$, for $k \geq 3$.

(b) Let n = -r, where r is a positive integer. Then, for $k \geq 1$, Equation (3) gives

$$\binom{-r}{k} = \frac{-r\cdot(-r-1)\cdots(-r-k+1)}{k!} = (-1)^k \binom{r+k-1}{k}.$$

Thus,

$$(1+x)^n = \frac{1}{(1+x)^r} = 1 - rx + \binom{r+1}{2}x^2 \sum_{k \geq 3}\binom{r+k-1}{k}(-x)^k.$$

The readers are advised to get the above expression using the Taylor series expansion of $(1+x)^n$ around $x = 0$.

2. Let $n, m \in \mathbb{N}$. Recall the identity $n^m = \sum_{k=0}^{m}\binom{n}{k}k!S(m, k) = \sum_{k=0}^{n}\binom{n}{k}k!S(m, k)$ that appeared in

Lemma: We note that for a fixed positive integer m, the above identity is same as the matrix product X = AY, where

$$X = \begin{bmatrix} 0^m \\ 1^m \\ 2^m \\ 3^m \\ \vdots \\ n^m \end{bmatrix}, \quad A = \begin{bmatrix} \binom{0}{0} & 0 & 0 & 0 & \cdots & 0 \\ \binom{1}{0} & \binom{1}{1} & 0 & 0 & \cdots & 0 \\ \binom{2}{0} & \binom{2}{1} & \binom{2}{2} & 0 & \cdots & 0 \\ \binom{3}{0} & \binom{3}{1} & \binom{3}{2} & \binom{3}{3} & \cdots & 0 \\ \vdots & \vdots & \vdots & \vdots & \ddots & \vdots \\ \binom{n}{0} & \binom{n}{1} & \binom{n}{2} & \binom{n}{3} & \cdots & \binom{n}{n} \end{bmatrix} \quad \text{and } Y = \begin{bmatrix} 0!S(m,0) \\ 1!S(m,1) \\ 2!S(m,2) \\ \vdots \\ n!S(m,n) \end{bmatrix}.$$

Hence, if we know the inverse of the matrix A, we can write Y = A⁻¹X. Check that

$$A^{-1} = \begin{bmatrix} \binom{0}{0} & 0 & 0 & 0 & \cdots & 0 \\ -\binom{1}{0} & \binom{1}{1} & 0 & 0 & \cdots & 0 \\ \binom{2}{0} & -\binom{2}{1} & \binom{2}{2} & 0 & \cdots & 0 \\ -\binom{3}{0} & \binom{3}{1} & -\binom{3}{2} & \binom{3}{3} & \cdots & 0 \\ \vdots & \vdots & \vdots & \vdots & \ddots & \vdots \\ (-1)^n\binom{n}{0} & (-1)^{n-1}\binom{n}{1} & (-1)^{n-2}\binom{n}{2} & (-1)^{n-3}\binom{n}{3} & \cdots & \binom{n}{n} \end{bmatrix}.$$

This gives us a way to calculate the Stirling numbers of second kind as a function of binomial co-efficients. That is, verify that

$$S(m,n) = \frac{1}{n!}\sum_{k\geq 0}(-1)^k \binom{n}{k}(n-k)^m.\qquad(4)$$

The above ideas imply that for all non-negative integers n the identity $a(n) = \sum_{k\geq 0}\binom{n}{k}b(k)$ holds if and only if $b(n) = \sum_{k\geq 0}(-1)^k\binom{n}{k}a(k)$ holds.

Example: On a rainy day, n students leave their umbrellas (which are indistinguishable) outside their examination room. Determine the number of ways of collecting the umbrellas so that no student collects the correct umbrella when they finish the examination? This problem is generally known by the Derangement problem.

Solution: Let the students be numbered 1, 2, . . . , n and suppose that the ith student has the umbrella numbered i, $1 \leq i \leq n$. So, we are interested in the number of permutations of the set $\{1, 2, . . . , n\}$ such that the number i is not at the ith position, for $1 \leq i \leq n$. Let D_n represent the number of derangements. Then it can be checked that $D_2 = 1$ and $D_3 = 2$. They correspond the permutations 21 for n = 2 and 231, 312 for n = 3. We will try to find a relationship of D_n with D_i, for $1 \leq i \leq n - 1$.

Let us have a close look at the required permutations. We note that n should not be placed at the nth position. So, n has to appear some where between 1 and n – 1. That is, for some i, $1 \leq i \leq n - 1$

(a)   n appears at the ith position and i appears at the nth position, or

(b)   n appears at the ith position and i does not appear at the nth position.

Case (a): Fix i, $1 \leq i \leq n - 1$. Then the position of n and i is fixed and the remaining numbers j, for $j \neq i, n$ should not appear at the jth place. As $j \neq i, n$, the problem reduces to the number of de-rangements of n – 2 numbers which by our notation equals D_{n-2}. As i can be any one of the integers 1, 2, . . . , n – 1, the number of derangements corresponding to the first case equals $(n - 1)D_{n-2}$.

Case (b): Fix i, $1 \leq i \leq n - 1$. Then the position of n is at the ith place but i is not placed at the nth position. So, in this case, the problem reduces to placing the numbers 1, 2, . . . , n – 1 at the places 1, 2, . . . , i – 1, i + 1, . . . , n such that the number i is not to be placed at the nth position and for j $j \neq i, j$ is not placed at the jth position. Let us rename the positions as $a_1, a_2, . . . , a_{n-1}$, where $a_i = n$ and $a_j = j$ for $j \neq i$.

Then, with this renaming, the problem reduces to placing the numbers 1, 2, . . . , n – 1 at places a_1, $a_2, . . . , a_{n-1}$ such that the number j, for $1 \leq j \leq n - 1$, is not placed at aj'^{th} position. This corresponds to the derangement of n – 1 numbers that this by our notation equals D_{n-1}. Thus, in this case the number of derangements equals $(n - 1)D_{n-1}$.

Hence, $D_n = (n - 1)D_{n-1} + (n - 1)D_{n-2}$. Or equivalently, $D_2 = 1$ and $D_1 = 0$ imply $D_n - nD_{n-1} = -(D_{n-1} - (n - 1)D_{n-2}) = (-1)^2(D_{n-2} - (n - 2)D_{n-3}) = \bullet\bullet\bullet = (-1)^n$.

Therefore,

$$D_n = nD_{n-1} + (-1)^n = n\left((n-1)D_{n-2} + (-1)^{n-1}\right) + (-1)^n$$

$$= n(n-1)D_{n-2} + n(-1)^{n-1} + (-1)^n$$

$$\vdots$$

$$= n(n-1)\cdots 4\cdot 3\, D_2 + n(n-1)\cdots 4(-1)^3 + \cdots + n(-1)^{n-1} + (-1)^n$$

$$= n!\left(1 + \frac{-1}{1!} + \frac{(-1)^2}{2!} + \cdots + \frac{(-1)^{n-1}}{(n-1)!} + \frac{(-1)^n}{n!}\right).$$

Or, in other words $\lim\limits_{n\to\infty} \dfrac{D_n}{n!} = \dfrac{1}{e}$.

References

- Bourbaki, N. (18 November 1998). Elements of the History of Mathematics Paperback. J. Meldrum (Translator). ISBN 978-3-540-64767-6

- Coolidge, J. L. (1949). "The Story of the Binomial Theorem". The American Mathematical Monthly. 56 (3): 147–157. doi:10.2307/2305028

- Landau, James A. (1999-05-08). "Historia Matematica Mailing List Archive: Re: [HM] Pascal's Triangle" (mailing list email). Archives of Historia Matematica. Retrieved 2007-04-13

- Cover, Thomas M.; Thomas, Joy A. (2001-01-01). Data Compression. John Wiley & Sons, Inc. p. 320. ISBN 9780471200611. doi:10.1002/0471200611.ch5

- Biggs, N. L. (1979). "The roots of combinatorics". Historia Math. 6 (2): 109–136. doi:10.1016/0315-0860(79)90074-0

Combinatorics and Generating Functions

Combinatorics deal with the study of finite or countable sets of discrete structures while generating functions deal with an infinite series of numbers. Pigeonhole principle, recurrence relation, formal power series, etc. are some of the main theories in this field. Discrete mathematics is best understood in confluence with the major topics listed in the following chapter.

Pigeonhole Principle

Pigeons in holes. Here there are $n = 10$ pigeons in $m = 9$ holes. Since 10 is greater than 9, the pigeonhole principle says that at least one hole has more than one pigeon.

In mathematics, the pigeonhole principle states that if n items are put into m containers, with $n > m > 0$, then at least one container must contain more than one item. This theorem is exemplified in real life by truisms like "there must be at least two left gloves or two right gloves in a group of three gloves". It is an example of a counting argument, and despite seeming intuitive it can be used to demonstrate possibly unexpected results; for example, that two people in London have the same number of hairs on their heads.

The first formalization of the idea is believed to have been made by Peter Gustav Lejeune Dirichlet in 1834 under the name *Schubfachprinzip* ("drawer principle" or "shelf principle"). For this reason it is also commonly called Dirichlet's box principle or Dirichlet's drawer principle. This should not be confused with Dirichlet's principle, a term introduced by Riemann that refers to the minimum principle for harmonic functions.

The principle has several generalizations and can be stated in various ways. In a more quantified version: for natural numbers k and m, if $n = km + 1$ objects are distributed among m sets, then the pigeonhole principle asserts that at least one of the sets will contain at least $k + 1$ objects. For arbitrary n and m this generalizes to $k + 1 = \lfloor (n - 1)/m \rfloor + 1$, where $\lfloor \ldots \rfloor$ is the floor function.

Though the most straightforward application is to finite sets (such as pigeons and boxes), it is also used with infinite sets that cannot be put into one-to-one correspondence. To do so requires the formal statement of the pigeonhole principle, which is *"there does not exist an injective function whose codomain is smaller than its domain"*. Advanced mathematical proofs like Siegel's lemma build upon this more general concept.

Etymology

P.G.L. Dirichlet published his works in both French and German. The strict original meaning of either the German *Schubfach*, or the French *tiroir*, corresponds to the English *drawer*, an *open-topped box that can be slid in and out of the cabinet that contains it*. These terms were morphed to the word *pigeonhole*, standing for a *small open space in a desk, cabinet, or wall for keeping letters or papers*, metaphorically rooted in the structures that house pigeons. Considering the fact that Dirichlet's father was a postmaster, necessarily best acquainted to furniture of type *pigeonhole*, common for sorting letters in his business, the translation by *pigeonholes* may be a perfect transfer of Dirichlet's terms of understanding. The meaning, referring to some furniture features, has since been strongly overtaken and is fading, especially among those who do not speak English natively, but as a lingua franca in the scientific world, in favour of the more pictorial interpretation, literally involving pigeons and holes. It is interesting to note that the suggestive, though not misleading interpretation of "pigeonhole" as "dovecote" has lately found its way back to a German "re-translation" of the "pigeonhole"-principle as the "Taubenschlag"-principle.

Examples

Sock-picking

Assume a drawer contains a mixture of black socks and blue socks, each of which can be worn on either foot, and that you are pulling a number of socks from the drawer without looking. What is the minimum number of pulled socks required to guarantee a pair of the same color? Using the pigeonhole principle, to have at least one pair of the same color ($m = 2$ holes, one per color) using one pigeonhole per color, you need to pull only three socks from the drawer ($n = 3$ items). Either you have *three* of one color, or, exclusively, *two* of one color and *one* of the other.

Hand-shaking

If there are n people who can shake hands with one another (where $n > 1$), the pigeonhole principle shows that there is always a pair of people who will shake hands with the same number of people. As the 'holes', or m, correspond to number of hands shaken, and each person can shake hands with anybody from 0 to $n - 1$ other people, this creates $n - 1$ possible holes. This is because either the '0' or the '$n - 1$' hole must be empty (if one person shakes hands with everybody, it's not possible to have another person who shakes hands with nobody; likewise, if one person shakes hands with no one there cannot be a person who shakes hands with everybody). This leaves n people to be placed in at most $n - 1$ non-empty holes, guaranteeing duplication.

Hair-counting

We can demonstrate there must be at least two people in London with the same number of hairs on their heads as follows. Since a typical human head has an average of around 150,000 hairs, it is reasonable to assume (as an upper bound) that no one has more than 1,000,000 hairs on their head (m = 1 million holes). There are more than 1,000,000 people in London (n is bigger than 1 million items). Assigning a pigeonhole to each number of hairs on a person's head, and assign people to pigeonholes according to the number of hairs on their head, there must be at least two people assigned to the same pigeonhole by the 1,000,001st assignment (because they have the same number of hairs on their heads) (or, $n > m$). For the average case (m = 150,000) with the constraint: fewest overlaps, there will be at most one person assigned to every pigeonhole and the 150,001st person assigned to the same pigeonhole as someone else. In the absence of this constraint, there may be empty pigeonholes because the "collision" happens before we get to the 150,001st person. The principle just proves the existence of an overlap; it says nothing of the number of overlaps (which falls under the subject of probability distribution).

There is a passing, satirical, allusion in English to this version of the principle in *A History of the Athenian Society*, prefixed to ""A Supplement to the Athenian Oracle: Being a Collection of the Remaining Questions and Answers in the Old Athenian Mercuries"", (Printed for Andrew Bell, London, 1710). It seems that the question *whether there were any two persons in the World that have an equal number of hairs on their head?* had been raised in *The Athenian Mercury* before 1704.

Perhaps the first written reference to the pigeonhole principle appears in 1622 in a short sentence of the Latin work *Selectæ Propositiones*, by the French Jesuit Jean Leurechon, where he wrote "It is necessary that two men have the same number of hairs, écus, or other things, as each other."

The Birthday Problem

The birthday problem asks, for a set of n randomly chosen people, what is the probability that some pair of them will have the same birthday? By the pigeonhole principle, if there are 367 people in the room, we know that there is at least one pair who share the same birthday, as there are only 366 possible birthdays to choose from (including February 29, if present). The birthday "paradox" refers to the result that even if the group is as small as 23 individuals, there will still be a pair of people with the same birthday with a 50% probability. While at first glance this may seem surprising, it intuitively makes sense when considering that a comparison will actually be made between every possible pair of people rather than fixing one individual and comparing them solely to the rest of the group.

Softball Team

Imagine seven people who want to play softball (n = 7 items), with a limitation of only four softball teams (m = 4 holes) to choose from. The pigeonhole principle tells us that they cannot all play for different teams; there must be at least one team featuring at least two of the seven players:

$$\left\lfloor \frac{n-1}{m} \right\rfloor + 1 = \left\lfloor \frac{7-1}{4} \right\rfloor + 1 = \left\lfloor \frac{6}{4} \right\rfloor + 1 = 1 + 1 = 2$$

Subset Sum

Any subset of size six from the set S = {1,2,3,...,9} must contain two elements whose sum is 10. The pigeonholes will be labelled by the two element subsets {1,9}, {2,8}, {3,7}, {4,6} and the singleton

{5}, five pigeonholes in all. When the six "pigeons" (elements of the size six subset) are placed into these pigeonholes, each pigeon going into the pigeonhole that has it contained in its label, at least one of the pigeonholes labelled with a two element subset will have two pigeons in it.

Uses and Applications

The pigeonhole principle arises in computer science. For example, collisions are inevitable in a hash table because the number of possible keys exceeds the number of indices in the array. A hashing algorithm, no matter how clever, cannot avoid these collisions.

The principle can be used to prove that any lossless compression algorithm, provided it makes some inputs smaller (as the name compression suggests), will also make some other inputs larger. Otherwise, the set of all input sequences up to a given length L could be mapped to the (much) smaller set of all sequences of length less than L without collisions (because the compression is lossless), a possibility which the pigeonhole principle excludes.

A notable problem in mathematical analysis is, for a fixed irrational number a, to show that the set $\{[na]: n \text{ is an integer}\}$ of fractional parts is dense in $[0, 1]$. One finds that it is not easy to explicitly find integers n, m such that $|na - ma| < e$, where $e > 0$ is a small positive number and a is some arbitrary irrational number. But if one takes M such that $1/M < e$, by the pigeonhole principle there must be n_1, $n_2 \in \{1, 2, ..., M + 1\}$ such that $n_1 a$ and $n_2 a$ are in the same integer subdivision of size $1/M$ (there are only M such subdivisions between consecutive integers). In particular, we can find n_1, n_2 such that $n_1 a$ is in $(p + k/M, p + (k + 1)/M)$, and $n_2 a$ is in $(q + k/M, q + (k + 1)/M)$, for some p, q integers and k in $\{0, 1, ..., M - 1\}$. We can then easily verify that $(n_2 - n_1)a$ is in $(q - p - 1/M, q - p + 1/M)$. This implies that $[na] < 1/M < e$, where $n = n_2 - n_1$ or $n = n_1 - n_2$. This shows that 0 is a limit point of $\{[na]\}$. We can then use this fact to prove the case for p in $(0, 1]$: find n such that $[na] < 1/M < e$; then if $p \in (0, 1/M]$, we are done. Otherwise $p \in (j/M, (j + 1)/M]$, and by setting $k = \sup\{r \in N : r[na] < j/M\}$, one obtains $|[(k + 1)na] - p| < 1/M < e$.

Variants occurring in well known proofs: In the proof of the pumping lemma for regular languages, a version that mixes finite and infinite sets is used: If infinitely many objects are placed in finitely many boxes, then there exist two objects that share a box. In Fisk's solution of the Art gallery problem a sort of converse is used: If n objects are placed in k boxes, then there is a box containing at most n/k objects.

Alternate Formulations

The following are alternate formulations of the pigeonhole principle.

1. If n objects are distributed over m places, and if $n > m$, then some place receives at least two objects.

2. (equivalent formulation of 1) If n objects are distributed over n places in such a way that no place receives more than one object, then each place receives exactly one object.

3. If n objects are distributed over m places, and if $n < m$, then some place receives no object.

4. (equivalent formulation of 3) If n objects are distributed over n places in such a way that no place receives no object, then each place receives exactly one object.

Strong Form

Let $q_1, q_2, ..., q_n$ be positive integers. If

$$q_1 + q_2 + \cdots + q_n - n + 1$$

objects are distributed into n boxes, then either the first box contains at least q_1 objects, or the second box contains at least q_2 objects, ..., or the nth box contains at least q_n objects.

The simple form is obtained from this by taking $q_1 = q_2 = ... = q_n = 2$, which gives $n + 1$ objects. Taking $q_1 = q_2 = ... = q_n = r$ gives the more quantified version of the principle, namely:

Let n and r be positive integers. If $n(r - 1) + 1$ objects are distributed into n boxes, then at least one of the boxes contains r or more of the objects.

This can also be stated as, if k discrete objects are to be allocated to n containers, then at least one container must hold at least $\lceil k / n \rceil$ objects, where $\lceil\ \rceil$ is the ceiling function, denoting the smallest integer larger than or equal to x. Similarly, at least one container must hold no more than $\lfloor k / n \rfloor$ objects, where $\lfloor x \rfloor$ is the floor function, denoting the largest integer smaller than or equal to x.

Generalizations of the Pigeonhole Principle

A probabilistic generalization of the pigeonhole principle states that if n pigeons are randomly put into m pigeonholes with uniform probability $1/m$, then at least one pigeonhole will hold more than one pigeon with probability

$$1 - \frac{(m)_n}{m^n},$$

where $(m)_n$ is the falling factorial $m(m - 1)(m - 2)...(m - n + 1)$. For $n = 0$ and for $n = 1$ (and $m > 0$), that probability is zero; in other words, if there is just one pigeon, there cannot be a conflict. For $n > m$ (more pigeons than pigeonholes) it is one, in which case it coincides with the ordinary pigeonhole principle. But even if the number of pigeons does not exceed the number of pigeonholes ($n \leq m$), due to the random nature of the assignment of pigeons to pigeonholes there is often a substantial chance that clashes will occur. For example, if 2 pigeons are randomly assigned to 4 pigeonholes, there is a 25% chance that at least one pigeonhole will hold more than one pigeon; for 5 pigeons and 10 holes, that probability is 69.76%; and for 10 pigeons and 20 holes it is about 93.45%. If the number of holes stays fixed, there is always a greater probability of a pair when you add more pigeons. This problem is treated at much greater length in the birthday paradox.

A further probabilistic generalization is that when a real-valued random variable X has a finite mean $E(X)$, then the probability is nonzero that X is greater than or equal to $E(X)$, and similarly the probability is nonzero that X is less than or equal to $E(X)$. To see that this implies the standard pigeonhole principle, take any fixed arrangement of n pigeons into m holes and let X be the number of pigeons in a hole chosen uniformly at random. The mean of X is n/m, so if there are more pigeons than holes the mean is greater than one. Therefore, X is sometimes at least 2.

Infinite Sets

The pigeonhole principle can be extended to infinite sets by phrasing it in terms of cardinal numbers: if the cardinality of set A is greater than the cardinality of set B, then there is no injection from A to B. However, in this form the principle is tautological, since the meaning of the statement that the cardinality of set A is greater than the cardinality of set B is exactly that there is no injective map from A to B. However, adding at least one element to a finite set is sufficient to ensure that the cardinality increases.

Another way to phrase the pigeonhole principle for finite sets is similar to the principle that finite sets are Dedekind finite: Let A and B be finite sets. If there is a surjection from A to B that is not injective, then no surjection from A to B is injective. In fact no function of any kind from A to B is injective. This is not true for infinite sets: Consider the function on the natural numbers that sends 1 and 2 to 1, 3 and 4 to 2, 5 and 6 to 3, and so on.

There is a similar principle for infinite sets: If uncountably many pigeons are stuffed into countably many pigeonholes, there will exist at least one pigeonhole having uncountably many pigeons stuffed into it.

This principle is not a generalization of the pigeonhole principle for finite sets however: It is in general false for finite sets. In technical terms it says that if A and B are finite sets such that any surjective function from A to B is not injective, then there exists an element of b of B such that there exists a bijection between the preimage of b and A. This is a quite different statement, and is absurd for large finite cardinalities.

Quantum Mechanics

Yakir Aharonov et al. have presented arguments that the pigeonhole principle may be violated in quantum mechanics, and proposed interferometric experiments to test the pigeonhole principle in quantum mechanics.

The pigeonhole principle states that if there are n + 1 pigeons and n holes (boxes), then there is at least one hole (box) that contains two or more pigeons. It can be easily verified that the pigeonhole principle is equivalent to the following statements:

1. If m pigeons are put into m pigeonholes, there is an empty hole if and only if there's a hole with more than one pigeon.

2. If n pigeons are put into m pigeonholes, with n > m, then there is a hole with more than one pigeon.

3. For two finite sets A and B, there exists a one to one and onto function f : A → B if and only if |A| = |B|.

Remark: Recall that the expression $\lceil x \rceil$, called the ceiling function, is the smallest integer ℓ, such that $\ell \geq x$ and the expression $\lfloor x \rfloor$, called the floor function, is the largest integer k, such that $k \leq x$.

1. [Generalized Pigeonhole Principle] if there are n pigeons and m holes with n > m, then there is at least one hole that contains $\lceil \frac{n}{m} \rceil$ pigeons.

2. Dirichlet was the one who popularized this principle.

Example: Let a be an irrational number. Then prove that there exist infinitely many rational numbers $s = \dfrac{p}{q}$, such that $|a - s| < \dfrac{1}{q^2}$.

Proof. Let $N \in \mathbb{N}$. Without loss of generality, we assume that a > 0. By {α}, we will denote the fractional part of α. That is, {α} = α − ⌊α⌋.

Now, consider the fractional parts {0}, {a}, {2a}, . . . , {Na} of the first (N + 1) multiples of a and the N subintervals $\left[0, \dfrac{1}{N}\right), \left[\dfrac{1}{N}, \dfrac{2}{N}\right), ..., \left[\dfrac{N-1}{N}, 1\right)$ of [0, 1). Clearly {ka}, for k a positive integer, cannot be an integer as a is an irrational number. Thus, by the pigeonhole principle, two of the above fractional parts must fall into the same subinterval. That is, there exist integers u, v and w such that u > v but

$$\{ua\} \in \left[\frac{w}{N}, \frac{w+1}{N}\right) \ and \ \{va\} \in \left[\frac{w}{N}, \frac{w+1}{N}\right).$$

Thus, $\left|\{ua\} - \{va\}\right| < \dfrac{1}{N}$ and $\left|\{ua\} - \{va\}\right| = |(u-v)a - (\lfloor ua\rfloor - \lfloor va\rfloor)|$. Now, let $q = u - v$ and $p = \lfloor ua\rfloor - \lfloor va\rfloor$. Then $p, q \in \mathbb{Z}, q \neq 0$ and $|qa - p| < \dfrac{1}{N}$. Dividing by q, we get

$$\left|a - \frac{p}{q}\right| < \frac{1}{Nq} \leq \frac{1}{q^2} \ as \ 0 < q \leq N.$$

Therefore, we have found a rational number $\dfrac{p}{q}$ such that $\left|a - \dfrac{p}{q}\right| < \dfrac{1}{q^2}$. We will now show that the number of such pairs (p, q) is infinite.

On the contrary, assume that there are only a finite number of rational numbers, say $r_1, r_2, \ldots, r_M$ such that

$$r_i = \frac{p_i}{q_i}, \ for \ i = 1, ..., M \ and \ \left|a - r_i\right| < \frac{1}{q_i^2}.$$

Since a is an irrational number, none of the differences $|a - r_i|$, for i = 1, 2, . . . , M , will be exactly 0. Therefore, there exists an integer Q such that

$$\left|a - r_i\right| > \frac{1}{Q}, \ for \ all \ i = 1, 2, \ldots, M.$$

We now, apply our earlier argument to this Q. The argument gives the existence of a fraction

$r = \dfrac{p}{q}$ such that $\left|a - r\right| > \dfrac{1}{Q_q} < \dfrac{1}{Q} < \left|a - r_i\right|,$ *for* $1 \leq i \leq M$ Hence, $r \neq r_i$, for all $i = 1, 2, \ldots, M$.

On the other hand, we also have, $\left|a - r\right| < \dfrac{1}{q^2}$ contradicting the assumption that the fractions r_i, for $i = 1, 2, \ldots, M$, were all the fractions with this property.

Example: 1. Let $\{a_1, a_2, \ldots, a_{mn+1}\}$ be a sequence of distinct $mn + 1$ real numbers. Then prove that this sequence has a subsequence of either $(m + 1)$ numbers that is strictly increasing or $(n + 1)$ numbers that is strictly decreasing.

Observation: The statement is NOT TRUE if there are exactly mn numbers. For example, consider the sequence 4, 3, 2, 1, 8, 7, 6, 5, 12, 11, 10, 9 of $12 = 3 \times 4$ distinct numbers. This sequence neither has an increasing subsequence of 4 numbers nor a decreasing subsequence of 5 numbers. Also, observe that if we take any number different from 1, 2, $\ldots$, 12 and place it at any position in the above sequence then it can be verified that there is either an increasing subsequence of 4 numbers or a decreasing subsequence of 5 numbers. For example, if we place 7.5,

(a) "before the number 4" or "between the numbers 8 and 7" or "after the number 9", then there is a decreasing subsequence of length 5.

(b) "between the numbers 4 and 8" or between the numbers 7 and 9, then there is an increasing subsequence of length 4.

Proof. Let T be the given sequence. That is, $T = \{a_k\}_{k=1}^{mn+1}$ and define $\ell_i = \underset{s}{max}\{s :$ an increasing subsequence of length s exists starting with $a_i\}$.

Then there are $mn+1$ positive integers $\ell_1, \ell_2, \ldots, \ell_{mn+1}$. If there exists a j, $1 \leq j \leq mn+1$, such that $\ell_j \geq m + 1$, then by definition of ℓ_j, there exists an increasing sequence of length $m + 1$ starting with a_j and thus the result follows. So, on the contrary assume that $\ell_i \leq m$, for $1 \leq i \leq mn + 1$.

That is, we have $mn + 1$ numbers $(\ell_1, \ldots, \ell_{mn+1})$ and all of them have to be put in the boxes numbered $1, 2, \ldots, m$. So, by the generalized pigeonhole principle, there are at least $\left\lceil \dfrac{mn+1}{m} \right\rceil = n+1$ numbers (ℓ_i's) that lies in the same box. Therefore, let us assume that there exist numbers $1 \leq i_1 < i_2 < \cdots < i_{n+1} \leq mn+1$, such that

$$\ell_{i_1} = \ell_{i_2} = \cdots = \ell_{i_{n+1}}.$$

That is, the length of the largest increasing subsequences of T starting with the numbers $a_{i_1}, a_{i_2}, \ldots, a_{i_{n+1}}$ are all equal. We now claim that $a_{i_1} > a_{i_2} > \ldots > a_{i_{n+1}}$.

We will show that $a_{i_1} > a_{i_2}$. A similar argument will give the other inequalities and hence the proof of the claim. On the contrary, let if $a_{i_1} < a_{i_2}$ (recall that a_i's are distinct) and consider a largest increasing subsequence $a_{i_2} = \alpha_1 < \alpha_2 < \cdots < \alpha_{i_2}$ of T, starting with a_{i_2}, that has length ℓ_{i_2}. This subsequence with the assumption that $a_{i_1} < a_{i_2}$ gives an increasing subsequence

$$a_{i_1} < a_{i_2} = \alpha_1 < \alpha_2 < \cdots < \alpha_{l_{i_2}}$$

of T, starting with a_{i_1}, of length $\ell_{i_2} + 1$. So, by definition of ℓ_i's, $\ell_{i_1} \geq \ell_{i_2} + 1$. This gives a contradiction to the equality, ℓ_{i_1} ℓ_{i_2}, in earlier Equation. Hence the proof of the example is complete.

2. Prove that there exist two powers of 3 whose difference is divisible by 2011.

Proof. Consider the set S = {1 = 3^0, 3, 3^2, 3^3, . . . , 3^{2011}}. Then |S| = 2012. Also, we know that when we divide positive integers by 2011 then the possible remainders are 0, 1, 2, . . . , 2010 (corresponding to exactly 2011 boxes). So, if we divide the numbers in S with 2011, then by pigeonhole principle there will exist at least two numbers $0 \leq i < j \leq 2011$, such that the remainders of 3^j and 3^i, when divided by 2011, are equal. That is, 2011 divides $3^j - 3^i$. Hence, this completes the proof.

Observe that this argument also implies that "there exists a positive integer ℓ such that 2011 divides $3^\ell - 1$" or "there exists a positive power of 3 that leaves a remainder 1 when divided by 2011" as gcd(3, 2011) = 1.

3. Prove that there exists a power of three that ends with 0001.

Proof. Consider the set S = {1 = 3^0, 3, 3^2, 3^3, . . .}. Now, let us divide each element of S by 10^4. As |S| > 10^4, there exist i > j such that the remainders of 3^i and 3^j, when are divided by 10^4, are equal. But gcd(10^4, 3) = 1 and thus, 10^4 divides $3^\ell - 1$. That is, $3^\ell - 1 = s \cdot 10^4$ for some positive integer s. That is, $3^\ell = s \cdot 10^4 + 1$ and hence the result follows.

Inclusion–exclusion Principle

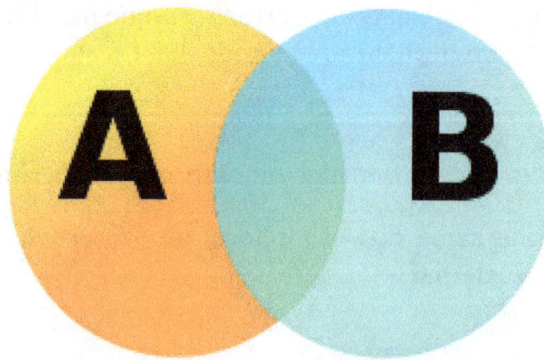

Venn diagram showing the union of sets A and B

In combinatorics (combinatorial mathematics), the inclusion–exclusion principle is a counting technique which generalizes the familiar method of obtaining the number of elements in the union of two finite sets; symbolically expressed as

$$|A \cup B| = |A| + |B| - |A \cap B|,$$

where A and B are two finite sets and $|S|$ indicates the cardinality of a set S (which may be considered as the number of elements of the set, if the set is finite). The formula expresses the fact that the sum of the sizes of the two sets may be too large since some elements may be counted twice. The double-counted elements are those in the intersection of the two sets and the count is corrected by subtracting the size of the intersection.

The principle is more clearly seen in the case of three sets, which for the sets A, B and C is given by

$$|A \cup B \cup C| = |A| + |B| + |C| - |A \cap B| - |A \cap C| - |B \cap C| + |A \cap B \cap C|.$$

This formula can be verified by counting how many times each region in the Venn diagram figure is included in the right-hand side of the formula. In this case, when removing the contributions of over-counted elements, the number of elements in the mutual intersection of the three sets has been subtracted too often, so must be added back in to get the correct total.

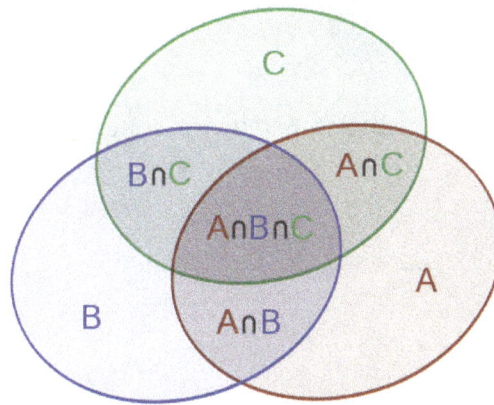

Inclusion–exclusion illustrated by a Venn diagram for three sets

Generalizing the results of these examples gives the principle of inclusion–exclusion. To find the cardinality of the union of n sets:

1. Include the cardinalities of the sets.

2. Exclude the cardinalities of the pairwise intersections.

3. Include the cardinalities of the triple-wise intersections.

4. Exclude the cardinalities of the quadruple-wise intersections.

5. Include the cardinalities of the quintuple-wise intersections.

6. Continue, until the cardinality of the n-tuple-wise intersection is included (if n is odd) or excluded (n even).

The name comes from the idea that the principle is based on over-generous *inclusion*, followed by compensating *exclusion*. This concept is attributed to Abraham de Moivre (1718); but it first appears in a paper of Daniel da Silva (1854), and later in a paper by J. J. Sylvester (1883). Sometimes the principle is referred to as the formula of Da Silva, or Sylvester due to these publications. The principle is an example of the sieve method extensively used in number theory and is sometimes referred to as the *sieve formula*, though Legendre already used a similar device in a sieve context in 1808.

As finite probabilities are computed as counts relative to the cardinality of the probability space, the formulas for the principle of inclusion–exclusion remain valid when the cardinalities of the sets are replaced by finite probabilities. More generally, both versions of the principle can be put under the common umbrella of measure theory.

In a very abstract setting, the principle of inclusion–exclusion can be expressed as the calculation of the inverse of a certain matrix. This inverse has a special structure, making the principle an extremely valuable technique in combinatorics and related areas of mathematics. As Gian-Carlo Rota put it:

"One of the most useful principles of enumeration in discrete probability and combinatorial theory is the celebrated principle of inclusion–exclusion. When skillfully applied, this principle has yielded the solution to many a combinatorial problem."

Statement

In its general form, the principle of inclusion–exclusion states that for finite sets $A_1, ..., A_n$, one has the identity

$$\left| \bigcup_{i=1}^{n} A_i \right| = \sum_{i=1}^{n} |A_i| - \sum_{1 \leq i < j \leq n} |A_i \cap A_j| + \cdots$$
$$\cdots + \sum_{1 \leq i < j < k \leq n} |A_i \cap A_j \cap A_k| - \cdots + (-1)^{n-1} |A_1 \cap \cdots \cap A_n|.$$

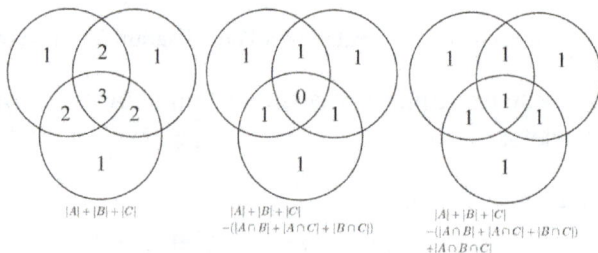

$$|A| + |B| + |C|$$

$$|A| + |B| + |C| \\ -(|A \cap B| + |A \cap C| + |B \cap C|)$$

$$|A| + |B| + |C| \\ -(|A \cap B| + |A \cap C| + |B \cap C|) \\ +|A \cap B \cap C|$$

Each term of the inclusion–exclusion formula gradually corrects the count until finally each portion of the Venn diagram is counted exactly once.

This can be compactly written as

$$\left| \bigcup_{i=1}^{n} A_i \right| = \sum_{k=1}^{n} (-1)^{k+1} \left(\sum_{1 \leq i_1 < \cdots < i_k \leq n} |A_{i_1} \cap \cdots \cap A_{i_k}| \right)$$

or

$$\left| \bigcup_{i=1}^{n} A_i \right| = \sum_{\emptyset \neq J \subseteq \{1,2,\ldots,n\}} (-1)^{|J|-1} \left| \bigcap_{j \in J} A_j \right|.$$

In words, to count the number of elements in a finite union of finite sets, first sum the cardinalities of the individual sets, then subtract the number of elements which appear in more than one set, then add back the number of elements which appear in more than two sets, then subtract the number of elements which appear in more than three sets, and so on. This process naturally ends since there can be no elements which appear in more than the number of sets in the union.

In applications it is common to see the principle expressed in its complementary form. That is, letting S be a finite universal set containing all of the A_i and letting $\overline{A_i}$ denote the complement of A_i in S, by De Morgan's laws we have

$$\left| \bigcap_{i=1}^{n} \overline{A_i} \right| = \left| S - \bigcup_{i=1}^{n} A_i \right|$$

$$= |S| - \sum_{i=1}^{n} |A_i| + \sum_{1 \le i < j \le n} |A_i \cap A_j| - \cdots + (-1)^n |A_1 \cap \cdots \cap A_n|.$$

As another variant of the statement, let P_1, P_2, ..., P_n be a list of properties that elements of a set S may or may not have, then the principle of inclusion–exclusion provides a way to calculate the number of elements of S which have none of the properties. Just let A_i be the subset of elements of S which have the property P_i and use the principle in its complementary form. This variant is due to J. J. Sylvester.

Examples

Counting Integers

As a simple example of the use of the principle of inclusion–exclusion, consider the question:

How many integers in {1,...,100} are not divisible by 2, 3 or 5?

Let S = {1,...,100} and P_1 the property that an integer is divisible by 2, P_2 the property that an integer is divisible by 3 and P_3 the property that an integer is divisible by 5. Letting A_i be the subset of S whose elements have property P_i we have by elementary counting: $|A_1|$ = 50, $|A_2|$ = 33, and $|A_3|$ = 20. There are 16 of these integers divisible by 6, 10 divisible by 10 and 6 divisible by 15. Finally, there are just 3 integers divisible by 30, so the number of integers not divisible by any of 2, 3 or 5 is given by:

$$100 - (50 + 33 + 20) + (16 + 10 + 6) - 3 = 26.$$

Counting Derangements

A more complex example is the following:

Suppose there is a deck of n cards numbered from 1 to n. Suppose a card numbered m is in the correct position if it is the mth card in the deck. How many ways, W, can the cards be shuffled with at least 1 card being in the correct position?

Begin by defining set A_m, which is all of the orderings of cards with the mth card correct. Then the number of orders, W, with *at least* one card being in the correct position, m, is

$$W = \left| \bigcup_{m=1}^{n} A_m \right|.$$

Apply the principle of inclusion–exclusion,

$$W = \sum_{m_1=1}^{n} |A_{m_1}| - \sum_{1 \le m_1 < m_2 \le n} |A_{m_1} \cap A_{m_2}| + \cdots$$

$$\cdots + \sum_{1 \le m_1 < m_2 < m_3 \le n} |A_{m_1} \cap A_{m_2} \cap A_{m_3}| - \cdots$$

$$\cdots + (-1)^{p-1} \sum_{1 \le m_1 < \cdots < m_p \le n} |A_{m_1} \cap \cdots \cap A_{m_p}| \cdots.$$

Each value $A_{m_1} \cap \cdots \cap A_{m_p}$ represents the set of shuffles having at least p values $m_1, \ldots, m_p$ in the correct position. Note that the number of shuffles with at least p values correct only depends on p, not on the particular values of m. For example, the number of shuffles having the 1st, 3rd, and 17th cards in the correct position is the same as the number of shuffles having the 2nd, 5th, and 13th cards in the correct positions. It only matters that of the n cards, 3 were chosen to be in the correct position. Thus there are $\binom{n}{p}$ equal terms in the pth summation.

$$W = \binom{n}{1}|A_1| - \binom{n}{2}|A_1 \cap A_2| + \binom{n}{3}|A_1 \cap A_2 \cap A_3| - \cdots$$

$$\cdots + (-1)^{p-1} \binom{n}{p} |A_1 \cap \cdots \cap A_p| \pm \cdots.$$

$|A_1 \cap \cdots \cap A_p|$ is the number of orderings having p elements in the correct position, which is equal to the number of ways of ordering the remaining $n - p$ elements, or $(n - p)!$. Thus we finally get:

$$W = \binom{n}{1}(n-1)! - \binom{n}{2}(n-2)! + \binom{n}{3}(n-3)! - \cdots$$

$$\cdots + (-1)^{p-1} \binom{n}{p}(n-p)! \pm \cdots.$$

$$W = \sum_{p=1}^{n} (-1)^{p-1} \binom{n}{p}(n-p)!.$$

Noting that $\binom{n}{p} = \dfrac{n!}{p!(n-p)!}$, this reduces to

$$W = \sum_{p=1}^{n} (-1)^{p-1} \frac{n!}{p!}.$$

A permutation where *no* card is in the correct position is called a derangement. Taking $n!$ to be the total number of permutations, the probability Q that a random shuffle produces a derangement is given by

$$Q = 1 - \frac{W}{n!} = \sum_{p=0}^{n} \frac{(-1)^p}{p!},$$

a truncation to $n + 1$ terms of the Taylor expansion of e^{-1}. Thus the probability of guessing an order for a shuffled deck of cards and being incorrect about every card is approximately $1/e$ or 37%.

A Special Case

The situation that appears in the derangement example above occurs often enough to merit special attention. Namely, when the size of the intersection sets appearing in the formulas for the principle of inclusion–exclusion depend only on the number of sets in the intersections and not on which sets appear. More formally, if the intersection

$$A_J := \bigcap_{j \in J} A_j$$

has the same cardinality, say $\alpha_k = |A_J|$, for every k-element subset J of $\{1, \ldots, n\}$, then

$$\left| \bigcup_{i=1}^{n} A_i \right| = \sum_{k=1}^{n} (-1)^{k-1} \binom{n}{k} \alpha_k.$$

Or, in the complementary form, where the universal set S has cardinality α_0,

$$\left| S \setminus \bigcup_{i=1}^{n} A_i \right| = \sum_{k=0}^{n} (-1)^{k} \binom{n}{k} \alpha_k.$$

A Generalization

Given a family (repeats allowed) of subsets $A_1, A_2, \ldots, A_n$ of a universal set S, the principle of inclusion–exclusion calculates the number of elements of S in none of these subsets. A generalization of this concept would calculate the number of elements of S which appear in exactly some fixed m of these sets.

Let $N = [n] = \{1, 2, \ldots, n\}$. If we define $A_\varnothing = S$, then the principle of inclusion–exclusion can be written as, using the notation of the previous section; the number of elements of S contained in none of the A_i is:

$$\sum_{J \subseteq [n]} (-1)^{|J|} |A_J|_.,$$

If I is a fixed subset of the index set N, then the number of elements which belong to A_i for all i in I and for no other values is:

$$\sum_{I \subseteq J} (-1)^{|J|-|I|} |A_J|.$$

Define the sets

$$B_k = A_{I \cup \{k\}} \text{ for } k \in N \setminus I.$$

We seek the number of elements in none of the B_k which, by the principle of inclusion–exclusion (with $B_\varnothing = A_I$), is

$$\sum_{K \subseteq N \setminus I} (-1)^{|K|} |B_K|.$$

The correspondence $K \leftrightarrow J = I \cup K$ between subsets of $N \setminus I$ and subsets of N containing I is a bijection and if J and K correspond under this map then $B_K = A_J$, showing that the result is valid.

In Probability

In probability, for events $A_1, ..., A_n$ in a probability space $(\Omega, \mathcal{F}, \mathbb{P})$, , the inclusion–exclusion principle becomes for $n = 2$

$$\mathbb{P}(A_1 \cup A_2) = \mathbb{P}(A_1) + \mathbb{P}(A_2) - \mathbb{P}(A_1 \cap A_2),$$

for $n = 3$

$$\mathbb{P}(A_1 \cup A_2 \cup A_3) = \mathbb{P}(A_1) + \mathbb{P}(A_2) + \mathbb{P}(A_3) \quad - \mathbb{P}(A_1 \cap A_2) - \mathbb{P}(A_1 \cap A_3) - \mathbb{P}(A_2 \cap A_3) \quad + \mathbb{P}(A_1 \cap A_2 \cap A_3)$$

and in general

$$\mathbb{P}\left(\bigcup_{i=1}^n A_i\right) = \sum_{i=1}^n \mathbb{P}(A_i) - \sum_{i<j} \mathbb{P}(A_i \cap A_j) \quad + \sum_{i<j<k} \mathbb{P}(A_i \cap A_j \cap A_k) - \cdots + (-1)^{n-1} \mathbb{P}\left(\bigcap_{i=1}^n A_i\right),$$

which can be written in closed form as

$$\mathbb{P}\left(\bigcup_{i=1}^n A_i\right) = \sum_{k=1}^n \left((-1)^{k-1} \sum_{\substack{I \subset \{1,...,n\} \\ |I|=k}} \mathbb{P}(A_I) \right),$$

where the last sum runs over all subsets I of the indices $1, ..., n$ which contain exactly k elements, and

$$A_I := \bigcap_{i \in I} A_i$$

denotes the intersection of all those A_i with index in I.

According to the Bonferroni inequalities, the sum of the first terms in the formula is alternately an upper bound and a lower bound for the LHS. This can be used in cases where the full formula is too cumbersome.

For a general measure space (S, Σ, μ) and measurable subsets $A_1, ..., A_n$ of finite measure, the above identities also hold when the probability measure $\mathbb{P}$ is replaced by the measure μ.

Special Case

If, in the probabilistic version of the inclusion–exclusion principle, the probability of the intersection A_I only depends on the cardinality of I, meaning that for every k in $\{1, ..., n\}$ there is an a_k such that

$$a_k = \mathbb{P}(A_I) \text{ for every } I \subset \{1,\ldots,n\} \text{ with } |I| = k,$$

then the above formula simplifies to

$$\mathbb{P}\left(\bigcup_{i=1}^{n} A_i\right) = \sum_{k=1}^{n}(-1)^{k-1}\binom{n}{k}a_k$$

due to the combinatorial interpretation of the binomial coefficient $\binom{n}{k}$.

An analogous simplification is possible in the case of a general measure space (S,Σ,μ) and measurable subsets $A_1, ..., A_n$ of finite measure.

Other Forms

The principle is sometimes stated in the form that says that if

$$g(A) = \sum_{S:S \subseteq A} f(S)$$

then

$$f(A) = \sum_{S:S \subseteq A} (-1)^{|A|-|S|}g(S) \qquad (**)$$

The combinatorial and the probabilistic version of the inclusion–exclusion principle are instances of (**).

Proof

Take $\underline{m} = \{1, 2, \ldots, m\}$, $f(\underline{m}) = 0$, and

$$f(S) = \left|\bigcap_{i \in \underline{m}\setminus S} A_i \setminus \bigcup_{i \in S} A_i\right| \text{ and } f(S) = \mathbb{P}\left(\bigcap_{i \in \underline{m}\setminus S} A_i \setminus \bigcup_{i \in S} A_i\right)$$

respectively for all sets S with $S \subsetneq \underline{m}$. Then we obtain

$$g(A) = \left|\bigcap_{i \in \underline{m}\setminus A} A_i\right|, \quad g(\underline{m}) = \left|\bigcup_{i \in \underline{m}} A_i\right| \text{ and } g(A) = \mathbb{P}\left(\bigcap_{i \in \underline{m}\setminus A} A_i\right), \quad g(\underline{m}) = \mathbb{P}\left(\bigcup_{i \in \underline{m}} A_i\right)$$

respectively for all sets A with $A \subsetneq \underline{m}$. This is because elements a of $\bigcap_{i \in \underline{m}\setminus A} A_i$ can be contained in other A_i's (A_i's with $i \in A$) as well, and the $\cap\setminus\cup$ formula runs exactly through all possible extensions of the sets $\{A_i \mid i \in \underline{m} \setminus A\}$ with other A_i's, counting a only for the set that matches the membership behavior of a, if S runs through all subsets of A (as in the definition of $g(A)$).

Since $f(\underline{m}) = 0$, we obtain from (**) with $A = \underline{m}$ that

$$\sum_{\underline{m} \supseteq T \supsetneq \varnothing} (-1)^{|T|-1} g(\underline{m} \setminus T) = \sum_{\varnothing \subseteq S \subsetneq \underline{m}} (-1)^{m-|S|-1} g(S) = g(\underline{m})$$

and by interchanging sides, the combinatorial and the probabilistic version of the inclusion–exclusion principle follow.

If one sees a number n as a set of its prime factors, then (**) is a generalization of Möbius inversion formula for square-free natural numbers. Therefore, (**) is seen as the Möbius inversion formula for the incidence algebra of the partially ordered set of all subsets of A.

For a generalization of the full version of Möbius inversion formula, (**) must be generalized to multisets. For multisets instead of sets, (**) becomes

$$f(A) = \sum_{S \subseteq A} \mu(A - S) g(S) \qquad (\text{***})$$

where $A - S$ is the multiset for which $(A - S) \uplus S = A$, and

- $\mu(S) = 1$ if S is a set (i.e. a multiset without double elements) of even cardinality.

- $\mu(S) = -1$ if S is a set (i.e. a multiset without double elements) of odd cardinality.

- $\mu(S) = 0$ if S is a proper multiset (i.e. S has double elements).

Notice that $\mu(A - S)$ is just the $(-1)^{|A|-|S|}$ of (**) in case $A - S$ is a set.

Proof of (***)

Substitute

$$g(S) = \sum_{T \subseteq S} f(T)$$

on the right hand side of (***). Notice that $f(A)$ appears once on both sides of (***). So we must show that for all T with $T \subsetneq A$, the terms $f(T)$ cancel out on the right hand side of (***). For that purpose, take a fixed T such that $T \subsetneq A$ and take an arbitrary fixed $a \in A$ such that $a \notin T$.

Notice that $A - S$ must be a set for each positive or negative appearance of $f(T)$ on the right hand side of (***) that is obtained by way of the multiset S such that $T \subseteq S \subseteq A$. Now each appearance of $f(T)$ on the right hand side of (***) that is obtained by way of S such that $A - S$ is a set that contains a cancels out with the one that is obtained by way of the corresponding S such that $A - S$ is a set that does not contain a. This gives the desired result.

Applications

The inclusion–exclusion principle is widely used and only a few of its applications can be mentioned here.

Counting Derangements

A well-known application of the inclusion–exclusion principle is to the combinatorial problem of counting all derangements of a finite set. A *derangement* of a set A is a bijection from A into itself that has no fixed points. Via the inclusion–exclusion principle one can show that if the cardinality of A is n, then the number of derangements is $[n! \, / \, e]$ where $[x]$ denotes the nearest integer to x.

The first occurrence of the problem of counting the number of derangements is in an early book on games of chance: *Essai d'analyse sur les jeux de hazard* by P. R. de Montmort (1678 – 1719) and was known as either "Montmort's problem" or by the name he gave it, "*problème des rencontres.*" The problem is also known as the *hatcheck problem*.

The number of derangements is also known as the subfactorial of n, written $!n$. It follows that if all bijections are assigned the same probability then the probability that a random bijection is a derangement quickly approaches $1/e$ as n grows.

Counting Intersections

The principle of inclusion–exclusion, combined with De Morgan's law, can be used to count the cardinality of the intersection of sets as well. Let $\overline{A_k}$ represent the complement of A_k with respect to some universal set A such that $A_k \subseteq A$ for each k. Then we have

$$\bigcap_{i=1}^{n} A_i = \overline{\bigcup_{i=1}^{n} \overline{A_i}}$$

thereby turning the problem of finding an intersection into the problem of finding a union.

Graph Coloring

The inclusion exclusion principle forms the basis of algorithms for a number of NP-hard graph partitioning problems, such as graph coloring.

A well known application of the principle is the construction of the chromatic polynomial of a graph.

Bipartite Graph Perfect Matchings

The number of perfect matchings of a bipartite graph can be calculated using the principle.

Number of Onto Functions

Given finite sets A and B, how many surjective functions (onto functions) are there from A to B? Without any loss of generality we may take $A = \{1,2,...,k\}$ and $B = \{1,2,...,n\}$, since only the cardinalities of the sets matter. By using S as the set of all functions from A to B, and defining, for each i in B, the property P_i as "the function misses the element i in B" (i is not in the image of the function), the principle of inclusion–exclusion gives the number of onto functions between A and B as:

$$\sum_{j=0}^{n}\binom{n}{j}(-1)^{j}(n-j)^{k}.$$

Permutations with Forbidden Positions

A permutation of the set $S = \{1,2,...,n\}$ where each element of S is restricted to not being in certain positions (here the permutation is considered as an ordering of the elements of S) is called a *permutation with forbidden positions*. For example, with $S = \{1,2,3,4\}$, the permutations with the restriction that the element 1 can not be in positions 1 or 3, and the element 2 can not be in position 4 are: 2134, 2143, 3124, 4123, 2341, 2431, 3241, 3421, 4231 and 4321. By letting A_i be the set of positions that the element i is not allowed to be in, and the property P_i to be the property that a permutation puts element i into a position in A_i, the principle of inclusion–exclusion can be used to count the number of permutations which satisfy all the restrictions.

In the given example, there are $12 = 2(3!)$ permutations with property P_1, $6 = 3!$ permutations with property P_2 and no permutations have properties P_3 or P_4 as there are no restrictions for these two elements. The number of permutations satisfying the restrictions is thus:

$$4! - (12 + 6 + 0 + 0) + (4) = 24 - 18 + 4 = 10.$$

The final 4 in this computation is the number of permutations having both properties P_1 and P_2. There are no other non-zero contributions to the formula.

Stirling Numbers of the Second Kind

The Stirling numbers of the second kind, $S(n,k)$ count the number of partitions of a set of n elements into k non-empty subsets (indistinguishable *boxes*). An explicit formula for them can be obtained by applying the principle of inclusion–exclusion to a very closely related problem, namely, counting the number of partitions of an n-set into k non-empty but distinguishable boxes (ordered non-empty subsets). Using the universal set consisting of all partitions of the n-set into k (possibly empty) distinguishable boxes, $A_1, A_2, ..., A_k$, and the properties P_i meaning that the partition has box A_i empty, the principle of inclusion–exclusion gives an answer for the related result. Dividing by $k!$ to remove the artificial ordering gives the Stirling number of the second kind:

$$S(n,k) = \frac{1}{k!}\sum_{t=0}^{k}(-1)^{t}\binom{k}{t}(k-t)^{n}.$$

Rook Polynomials

A rook polynomial is the generating function of the number of ways to place non-attacking rooks on a *board B* that looks like a subset of the squares of a checkerboard; that is, no two rooks may be in the same row or column. The board B is any subset of the squares of a rectangular board with n rows and m columns; we think of it as the squares in which one is allowed to put a rook. The coefficient, $r_k(B)$ of x^k in the rook polynomial $R_B(x)$ is the number of ways k rooks, none of which attacks another, can be arranged in the squares of B. For any board B, there is a complementary board B' consisting of the squares of the rectangular board that are not in B. This complementary board also has a rook polynomial $R_{B'}(x)$ with coefficients $r_k(B')$.

It is sometimes convenient to be able to calculate the highest coefficient of a rook polynomial in terms of the coefficients of the rook polynomial of the complementary board. Without loss of generality we can assume that $n \leq m$, so this coefficient is $r_n(B)$. The number of ways to place n non-attacking rooks on the complete $n \times m$ "checkerboard" (without regard as to whether the rooks are placed in the squares of the board B) is given by the falling factorial:

$$(m)_n = m(m-1)(m-2)\cdots(m-n+1).$$

Letting P_i be the property that an assignment of n non-attacking rooks on the complete board has a rook in column i which is not in a square of the board B, then by the principle of inclusion–exclusion we have:

$$r_n(B) = \sum_{t=0}^{n}(-1)^t(m-t)_{n-t}\, r_t(B').$$

Euler's Phi Function

Euler's totient or phi function, $\varphi(n)$ is an arithmetic function that counts the number of positive integers less than or equal to n that are relatively prime to n. That is, if n is a positive integer, then $\varphi(n)$ is the number of integers k in the range $1 \leq k \leq n$ which have no common factor with n other than 1. The principle of inclusion–exclusion is used to obtain a formula for $\varphi(n)$. Let S be the set $\{1,2,...,n\}$ and define the property P_i to be that a number in S is divisible by the prime number p_i, for $1 \leq i \leq r$, where the prime factorization of

$$n = p_1^{a_1} p_2^{a_2} \cdots p_r^{a_r}.$$

Then,

$$\varphi(n) = n - \sum_{i=1}^{r}\frac{n}{p_i} + \sum_{1\leq i<j\leq r}\frac{n}{p_i p_j} - \cdots = n\prod_{i=1}^{r}\left(1-\frac{1}{p_i}\right).$$

Diluted Inclusion–exclusion Principle

In many cases where the principle could give an exact formula (in particular, counting prime numbers using the sieve of Eratosthenes), the formula arising doesn't offer useful content because the number of terms in it is excessive. If each term individually can be estimated accurately, the accumulation of errors may imply that the inclusion–exclusion formula isn't directly applicable. In number theory, this difficulty was addressed by Viggo Brun. After a slow start, his ideas were taken up by others, and a large variety of sieve methods developed. These for example may try to find upper bounds for the "sieved" sets, rather than an exact formula.

Let $A_1, ..., A_n$ be arbitrary sets and $p_1, ..., p_n$ real numbers in the closed unit interval $[0,1]$. Then, for every even number k in $\{0, ..., n\}$, the indicator functions satisfy the inequality:

$$1_{A_1 \cup \cdots \cup A_n} \geq \sum_{j=1}^{k} (-1)^{j-1} \sum_{1 \leq i_1 < \cdots < i_j \leq n} p_{i_1} \cdots p_{i_j} 1_{A_{i_1} \cap \cdots \cap A_{i_j}}.$$

Proof of Main Statement

Choose an element contained in the union of all sets and let $A_1, A_2, \ldots, A_t$ be the individual sets containing it. (Note that $t > 0$.) Since the element is counted precisely once by the left-hand side of earlier equation, we need to show that it is counted precisely once by the right-hand side. On the right-hand side, the only non-zero contributions occur when all the subsets in a particular term contain the chosen element, that is, all the subsets are selected from $A_1, A_2, \ldots, A_t$. The contribution is one for each of these sets (plus or minus depending on the term) and therefore is just the (signed) number of these subsets used in the term. We then have:

$$|\{A_i \mid 1 \leq i \leq t\}| - |\{A_i \cap A_j \mid 1 \leq i < j \leq t\}| + \cdots$$

$$\cdots + (-1)^{t+1} |\{A_1 \cap A_2 \cap \cdots \cap A_t\}|$$

$$= \binom{t}{1} - \binom{t}{2} + \cdots + (-1)^{t+1} \binom{t}{t}.$$

By the binomial theorem,

$$0 = (1-1)^t = \binom{t}{0} - \binom{t}{1} + \binom{t}{2} - \cdots + (-1)^t \binom{t}{t}.$$

Using the fact that $\binom{t}{0} = 1$ and rearranging terms, we have

$$1 = \binom{t}{1} - \binom{t}{2} + \cdots + (-1)^{t+1} \binom{t}{t},$$

and so, the chosen element is counted only once by the earlier equation.

Algebraic Proof

An algebraic proof can be obtained using indicator functions (characteristic functions of subsets of a set). The indicator function of a subset S of a set X is a function

$$1_S : X \to \{0,1\}$$

defined as

$$1_S(x) := \begin{cases} 1 & \text{if } x \in S, \\ 0 & \text{if } x \notin S. \end{cases}$$

If A and B are two subsets of X, then

$$1_A \cdot 1_B = 1_{A \cap B}.$$

Let A denote the union $\bigcup_{i=1}^{n} A_i$ of the sets $A_1, \ldots, A_n$. To prove the inclusion–exclusion principle in general, we first verify the identity

$$1_A = \sum_{k=1}^{n} (-1)^{k-1} \sum_{\substack{I \subset \{1,\ldots,n\} \\ |I|=k}} 1_{A_I} \qquad (*)$$

for indicator functions, where

$$A_I = \bigcap_{i \in I} A_i.$$

The following function is identically zero

$$(1_A - 1_{A_1})(1_A - 1_{A_2}) \cdots (1_A - 1_{A_n}) = 0,$$

because: if x is not in A, then all factors are 0 − 0 = 0; and otherwise, if x does belong to some A_m, then the corresponding mth factor is 1 − 1 = 0. By expanding the product on the left-hand side, equation ($*$) follows.

To prove the inclusion–exclusion principle for the cardinality of sets, sum the equation ($*$) over all x in the union of $A_1, \ldots, A_n$. To derive the version used in probability, take the expectation in ($*$). In general, integrate the equation ($*$) with respect to μ. Always use linearity in these derivations.

Principle of Inclusion and Exclusion

The following result is well known and hence we omit the proof.

Theorem: Let U be a finite set. Suppose A and B are two subsets of U. Then the number of elements of U that are neither in A nor in B are

$$|U| - (|A| + |B| - |A \cap B|).$$

Or equivalently, $|A \cup B| = |A| + |B| - |A \cap B|$.

A generalization of this to three subsets A, B and C is also well known. To get a result that generalizes Theorem for n subsets $A_1, A_2, \ldots, A_n$, we need the following notations:

$$S_1 = \sum_{i=1}^{n} |A_i|, \, S_2 = \sum_{1 \le i < j \le n} |A_i \cap A_j|, \, S_3 = \sum_{1 \le i < j < k \le n} \left| A_i \cap A_j \cap A_k \right|, \cdots, \, S_n = \left| A_1 \cap A_2 \cap \cdots \cap A_n \right|.$$

With the notations as defined above, we have the following theorem, called the inclusion- exclusion principle. This theorem can be easily proven using the principle of mathematical induction. But we give a separate proof for better understanding.

Theorem: [Inclusion-Exclusion Principle] Let $A_1, A_2, \ldots, A_n$ be n subsets of a finite set U. T h e n the number of elements of U that are in none of $A_1, A_2, \ldots, A_n$ is given by

$$|U| - S_1 + S_2 - S_3 + \cdots + (-1)^n S_n.$$

(1)

Or equivalently,

$$|A_1 \cup A_2 \cup \cdots \cup A_n| = S_1 - S_2 + \cdots + (-1 \quad S_n.$$

(2)

Proof. We show that if an element $x \in U$ belongs to exactly k of the subsets $A_1, A_2, \ldots, A_n$, for some $k \geq 1$, then its contribution in (1) is zero. Suppose x belongs to exactly k subsets $A_{i_1}, A_{i_2}, \ldots, A_{i_k}$. Then we observe the following:

1. The contribution of x in $|U|$ $is 1$.

2. The contribution of x in S_1 is k as $x \in A_{i_j}, 1 \leq j \leq k$.

3. The contribution of x in S_2 is $\binom{k}{2}$ as $x \in A_{i_j} \cap A_{i_l}, 1 \leq j < l \leq k.$

4. The contribution of x in S_3 is $\binom{k}{3}$ as $x \in A_{i_j} \cap A_{i_l} \cap A_{i_m}, 1 \leq j < l < m \leq k.$

Proceeding this way, we have

5. The contribution of x in S_k is $\binom{k}{k} = 1$, and

6. The contribution of x in S_ℓ for $\ell \geq k+1$ is 0.

So, the contribution of x in (1) is

$$1 - k + \binom{k}{2} - \binom{k}{3} + \cdots + (-1)^{k-1} \binom{k}{k-1} + (-1)^k \binom{k}{k} + 0 \cdots + 0 = (1-1)^k = 0.$$

This completes the proof of the theorem. The readers are advised to prove the equivalent condition.

Example: 1. Determine the number of 10-letter words using ENGLISH alphabets that does not contain all the vowels.

Solution: Let U be the set consisting of all the 10-letters words using ENGLISH alphabets and let Aα be a subset of U that does not contain the letter α. Then we need to compute

$$|A_a \cup A_e \cup A_i \cup A_o \cup A_u| = S_1 - S_2 + S_3 - S_4 + S_5,$$

where $S_1 = \displaystyle\sum_{\alpha \in \{a,e,i,o,u\}} |A_\alpha| = \binom{5}{1} 25^{10}, S_2 = \binom{5}{2} 24^{10}, S_3 = \binom{5}{3} 23^{10}, S_4 = \binom{5}{4} 22^{10}$ and $S_5 = 21^{10}$. So, the required answer is $\displaystyle\sum_{k=1}^{5} (-1)^{k-1} \binom{5}{k} (26-k)^{10}$.

2. Determine the number of integers between 1 and 1000 that are coprime to 2, 3, 11 and 13.

Solution: Let U = {1, 2, 3, . . . , 1000} and let A_i = {n ∈ U : i divides n}, for i = 2, 3, 11, 13. Then note that we need the value of |U| − |$A_2 \cup A_3 \cup A_{11} \cup A_{13}$|. Observe that

$$|A_2| = \left\lfloor \frac{1000}{2} \right\rfloor = 500, |A_3| = \left\lfloor \frac{1000}{3} \right\rfloor = 333, |A_{11}| = \left\lfloor \frac{1000}{11} \right\rfloor = 90, |A_{13}| = \left\lfloor \frac{1000}{13} \right\rfloor = 76,$$

$$|A_2 \cap A_3| = \left\lfloor \frac{1000}{6} \right\rfloor = 166, |A_2 \cap A_{11}| = \left\lfloor \frac{1000}{22} \right\rfloor = 45, |A_2 \cap A_{13}| = \left\lfloor \frac{1000}{26} \right\rfloor = 38,$$

$$|A_3 \cap A_{11}| = \left\lfloor \frac{1000}{33} \right\rfloor = 30, |A_3 \cap A_{13}| = \left\lfloor \frac{1000}{39} \right\rfloor = 25, |A_{11} \cap A_{13}| = \left\lfloor \frac{1000}{143} \right\rfloor = 6,$$

$$|A_2 \cap A_3 \cap A_{11}| = 15, |A_2 \cap A_3 \cap A_{13}| = 12, |A_2 \cap A_{11} \cap A_{13}| = 3,$$

$$|A_3 \cap A_{11} \cap A_{13}| = 2, |A_2 \cap A_3 \cap A_{11} \cap A_{13}| = 1.$$

Thus, the required number is

1000−((500+333+90+76)−(166+45+38+30+25+6)−(15+12+3+2)−1) = 1000−720 = 280.

3. (Euler's ϕ-function Or Euler's totient function) Let n denote a positive integer. Then the Euler ϕ-function is defined by

$$\phi(n) = \left| \{ k : 1 \le k \le n, \gcd(n, k) = 1 \} \right|. \tag{3}$$

Determine a formula for $\phi(n)$ in terms of its prime factors.

Solution: Let $n = p_1^{\alpha 1} p_2^{\alpha 2} \cdots p_k^{\alpha k}$ be the unique decomposition of n as product of distinct primes $p_1, p_2, \ldots, p_k$, $U = \{1, 2, \ldots, n\}$ and let $A_{p_i} = \{m \in U : p_i \text{ divides } m\}$, for $1 \le i \le k$ Then, by definition

$$\phi(n) = |U| - S_1 + S_2 - S_3 + \cdots + (-1)^k S_k$$

$$= n - \sum_{i=1}^{k} \frac{n}{p_i} + \sum_{1 \le i < j \le k} \frac{n}{p_i p_j} - \cdots + (-1)^k \frac{n}{p_1 p_2 \cdots p_k}$$

$$= n \prod_{i=1}^{k} \left(1 - \frac{1}{p_i} \right) \tag{4}$$

Formal Power Series

In mathematics, a formal power series is a generalization of a polynomial, where the number of terms is allowed to be infinite; this implies giving up the possibility of replacing the variable in the polynomial with an arbitrary number. Thus a formal power series differs from a polynomial in that it may have infinitely many terms, and differs from a power series, whose variables can take on numerical values. One way to view a formal power series is as an infinite ordered sequence of numbers. In this case, the powers of the variable are used only to indicate the order of the coefficients, so that the coefficient of x^5 is the fifth term in the sequence. In combinatorics, formal power series provide representations of numerical sequences and of multisets, and for instance allow concise expressions for recursively defined sequences regardless of whether the recursion can be explicitly solved; this is known as the method of generating functions. More generally, formal power series can include series with any finite number of variables, and with coefficients in an arbitrary ring.

Introduction

A formal power series can be loosely thought of as an object that is like a polynomial, but with infinitely many terms. Alternatively, for those familiar with power series (or Taylor series), one may think of a formal power series as a power series in which we ignore questions of convergence by not assuming that the variable X denotes any numerical value (not even an unknown value). For example, consider the series

$$A = 1 - 3X + 5X^2 - 7X^3 + 9X^4 - 11X^5 + \cdots.$$

If we studied this as a power series, its properties would include, for example, that its radius of convergence is 1. However, as a formal power series, we may ignore this completely; all that is relevant is the sequence of coefficients [1, −3, 5, −7, 9, −11, ...]. In other words, a formal power series is an object that just records a sequence of coefficients. It is perfectly acceptable to consider a formal power series with the factorials [1, 1, 2, 6, 24, 120, 720, 5040, ...] as coefficients, even though the corresponding power series diverges for any nonzero value of X.

Arithmetic on formal power series is carried out by simply pretending that the series are polynomials. For example, if

$$B = 2X + 4X^3 + 6X^5 + \cdots,$$

then we add A and B term by term:

$$A + B = 1 - X + 5X^2 - 3X^3 + 9X^4 - 5X^5 + \cdots.$$

We can multiply formal power series, again just by treating them as polynomials

$$AB = 2X - 6X^2 + 14X^3 - 26X^4 + 44X^5 + \cdots.$$

Notice that each coefficient in the product AB only depends on a *finite* number of coefficients of A and B. For example, the X^5 term is given by

$$44X^5 = (1 \times 6X^5) + (5X^2 \times 4X^3) + (9X^4 \times 2X).$$

For this reason, one may multiply formal power series without worrying about the usual questions of absolute, conditional and uniform convergence which arise in dealing with power series in the setting of analysis.

Once we have defined multiplication for formal power series, we can define multiplicative inverses as follows. The multiplicative inverse of a formal power series A is a formal power series C such that $AC = 1$, provided that such a formal power series exists. It turns out that if A has a multiplicative inverse, it is unique, and we denote it by A^{-1}. Now we can define division of formal power series by defining B/A to be the product BA^{-1}, provided that the inverse of A exists. For example, one can use the definition of multiplication above to verify the familiar formula

$$\frac{1}{1+X} = 1 - X + X^2 - X^3 + X^4 - X^5 + \cdots.$$

An important operation on formal power series is coefficient extraction. In its most basic form, the coefficient extraction operator for a formal power series in one variable extracts the coefficient of $[X^n]$, and is written e.g. $[X^n]A$, so that $[X^2]A = 5$ and $[X^5]A = -11$. Other examples include

$$\begin{aligned}
\left[X^3\right](B) &= 4, \\
\left[X^2\right](X + 3X^2Y^3 + 10Y^6) &= 3Y^3, \\
\left[X^2Y^3\right](X + 3X^2Y^3 + 10Y^6) &= 3, \\
\left[X^n\right]\left(\frac{1}{1+X}\right) &= (-1)^n, \\
\left[X^n\right]\left(\frac{X}{(1-X)^2}\right) &= n.
\end{aligned}$$

Similarly, many other operations that are carried out on polynomials can be extended to the formal power series setting, as explained below.

The Ring of Formal Power Series

The set of all formal power series in X with coefficients in a commutative ring R form another ring that is written $R[[X]]$, and called the ring of formal power series in the variable X over R.

Definition of the Formal Power Series Ring

One can characterize $R[[X]]$ abstractly as the completion of the polynomial ring $R[X]$ equipped with a particular metric. This automatically gives $R[[X]]$ the structure of a topological ring (and even of a complete metric space). But the general construction of a completion of a metric space is more involved than what is needed here, and would make formal power series seem more complicated than they are. It is possible to describe $R[[X]]$ more explicitly, and define the ring structure and topological structure separately, as follows.

Ring Structure

As a set, $R[[X]]$ can be constructed as the set $R^{\mathbb{N}}$ of all infinite sequences of elements of R, indexed by the natural numbers (taken to include 0). Designating a sequence whose term at index n is a_n by (a_n), one defines addition of two such sequences by

$$(a_n)_{n \in \mathbb{N}} + (b_n)_{n \in \mathbb{N}} = (a_n + b_n)_{n \in \mathbb{N}}$$

and multiplication by

$$(a_n)_{n \in \mathbb{N}} \times (b_n)_{n \in \mathbb{N}} = \left(\sum_{k=0}^{n} a_k b_{n-k} \right)_{n \in \mathbb{N}}.$$

This type of product is called the Cauchy product of the two sequences of coefficients, and is a sort of discrete convolution. With these operations, $R^{\mathbb{N}}$ becomes a commutative ring with zero element $(0,0,0,\cdots)$ and multiplicative identity $(1,0,0,\cdots)$.

The product is in fact the same one used to define the product of polynomials in one indeterminate, which suggests using a similar notation. One embeds R into $R[[X]]$ by sending any (constant) $a \in R$ to the sequence $(a,0,0,\cdots)$ and designates the sequence $(0,1,0,0,\cdots)$ by X; then using the above definitions every sequence with only finitely many nonzero terms can be expressed in terms of these special elements as

$$(a_0, a_1, a_2, \cdots, a_n, 0, 0, \ldots) = a_0 + a_1 X + \cdots + a_n X^n = \sum_{i=0}^{n} a_i X^i;$$

these are precisely the polynomials in X. Given this, it is quite natural and convenient to designate a general sequence $(a_n)_{n \in \mathbb{N}}$ by the formal expression $\sum_{i \in \mathbb{N}} a_i X^i$, even though the latter *is not* an expression formed by the operations of addition and multiplication defined above (from which only finite sums can be constructed). This notational convention allows reformulation the above definitions as

$$\left(\sum_{i \in \mathbb{N}} a_i X^i \right) + \left(\sum_{i \in \mathbb{N}} b_i X^i \right) = \sum_{i \in \mathbb{N}} (a_i + b_i) X^i$$

and

$$\left(\sum_{i \in \mathbb{N}} a_i X^i \right) \times \left(\sum_{i \in \mathbb{N}} b_i X^i \right) = \sum_{n \in \mathbb{N}} \left(\sum_{k=0}^{n} a_k b_{n-k} \right) X^n.$$

which is quite convenient, but one must be aware of the distinction between formal summation (a mere convention) and actual addition.

Topological Structure

Having stipulated conventionally that

$$(a_0, a_1, a_2, a_3, \ldots) = \sum_{i=0}^{\infty} a_i X^i,$$

one would like to interpret the right hand side as a well-defined infinite summation. To that end, a notion of convergence in $R^{\mathbb{N}}$ is defined and a topology on $R^{\mathbb{N}}$ is constructed. There are several equivalent ways to define the desired topology.

- We may give $R^{\mathbb{N}}$ the product topology, where each copy of R is given the discrete topology.

- We may give $R^{\mathbb{N}}$ the I-adic topology, where $I = (X)$ is the ideal generated by X, which consists of all sequences whose first term a_0 is zero.

- The desired topology could also be derived from the following metric. The distance between distinct sequences (a_n) and (b_n) in $R^{\mathbb{N}}$, is defined to be

$$d((a_n),(b_n)) = 2^{-k},$$

where k is the smallest natural number such that $a_k \neq b_k$; the distance between two equal sequences is of course zero.

Informally, two sequences $\{a_n\}$ and $\{b_n\}$ become closer and closer if and only if more and more of their terms agree exactly. Formally, the sequence of partial sums of some infinite summation converges if for every fixed power of X the coefficient stabilizes: there is a point beyond which all further partial sums have the same coefficient. This is clearly the case for the right hand side, regardless of the values a_n, since inclusion of the term for $\{i = n\}$ gives the last (and in fact only) change to the coefficient of X^n. It is also obvious that the limit of the sequence of partial sums is equal to the left hand side.

This topological structure, together with the ring operations described above, form a topological ring. This is called the ring of formal power series over R and is denoted by $R[[X]]$. The topology has the useful property that an infinite summation converges if and only if the sequence of its terms converges to 0, which just means that any fixed power of X occurs in only finitely many terms.

The topological structure allows much more flexible use of infinite summations. For instance the rule for multiplication can be restated simply as

$$\left(\sum_{i\in\mathbb{N}} a_i X^i\right) \times \left(\sum_{i\in\mathbb{N}} b_i X^i\right) = \sum_{i,j\in\mathbb{N}} a_i b_j X^{i+j},$$

since only finitely many terms on the right affect any fixed X^n. Infinite products are also defined by the topological structure; it can be seen that an infinite product converges if and only if the sequence of its factors converges to 1.

Alternative Topologies

The above topology is the finest topology for which $\sum_{i=0}^{\infty} a_i X^i$ always converges as a summation to the formal power series designated by the same expression, and it often suffices to give a meaning to infinite sums and products, or other kinds of limits that one wishes to use to designate particular formal power series. It can however happen occasionally that one wishes to use a coarser topology, so that certain expressions become convergent that would otherwise diverge. This applies in particular when the base ring $\mathbb{R}$ already comes with a topology other than the discrete one, for instance if it is also a ring of formal power series.

Consider the ring of formal power series

$$\mathbb{Z}[[X]][[Y]]$$

then the topology of above construction only relates to the indeterminate Y, since the topology that was put on $\mathbb{Z}[[X]]$ has been replaced by the discrete topology when defining the topology of the whole ring. So

$$\sum_{i\in\mathbb{N}} XY^i$$

converges to the power series suggested, which can be written as $\dfrac{X}{1-Y}$; however the summation

$$\sum_{i\in\mathbb{N}} X^i Y$$

would be considered to be divergent, since every term affects the coefficient of Y (which coefficient is itself a power series in X). This asymmetry disappears if the power series ring in Y is given the product topology where each copy of $\mathbb{Z}[[X]]$ is given its topology as a ring of formal power series rather than the discrete topology. As a consequence, for convergence of a sequence of elements of $\mathbb{Z}[[X]][[Y]]$ it then suffices that the coefficient of each power of Y converges to a formal power series in X, a weaker condition that stabilizing entirely; for instance in the second example given here the coefficient of Y converges to $\dfrac{1}{1-X}$, so the whole summation converges to $\dfrac{Y}{1-X}$.

This way of defining the topology is in fact the standard one for repeated constructions of rings of formal power series, and gives the same topology as one would get by taking formal power series in all inderteminates at once. In the above example that would mean constructing $\mathbb{Z}[[X,Y]]$, and here a sequence converges if and only if the coefficient of every monomial $X^i Y^j$ stabilizes. This topology, which is also the I-adic topology, where $I=(X,Y)$ is the ideal generated by X and Y, still enjoys the property that a summation converges if and only if its terms tend to 0.

The same principle could be used to make other divergent limits converge. For instance in $\mathbb{R}[[X]]$ the limit

$$\lim_{n\to\infty}\left(1+\frac{X}{n}\right)^n$$

does not exist, so in particular it does not converge to $\exp(X)=\sum_{n\in\mathbb{N}}\dfrac{X^n}{n!}$. This is because for $i\geq 2$ the coefficient $\dbinom{n}{i}/n^i$ of X^i does not stabilize as $n\to\infty$. It does however converge in the usual topology of $\mathbb{R}$, and in fact to the coefficient $\dfrac{1}{i!}$ of $\exp(X)$. Therefore, if one would give $\mathbb{R}[[X]]$ the product topology of $\mathbb{R}^{\mathbb{N}}$ where the topology of $\mathbb{R}$ is the usual topology rather than the discrete one, then the above limit would converge to $\exp(X)$. This more permissive approach is not however the standard when considering formal power series, as it would lead to convergence considerations that are as subtle as they are in analysis, while the philosophy of formal power series is on the contrary to make convergence questions as trivial as they can possibly be. With this topology it would *not* be the case that a summation converges if and only if its terms tend to 0.

Universal Property

The ring $\mathbb{R}[[X]]$ may be characterized by the following universal property. If S is a commutative associative algebra over R, if I is an ideal of S such that the I-adic topology on S is complete, and if x is an element of I, then there is a *unique* $\Phi : R[[X]] \to S$ with the following properties:

- Φ is an R-algebra homomorphism

- Φ is continuous

- $\Phi(X) = x$.

Operations on Formal Power Series

One can perform algebraic operations on power series to generate new power series. Besides the ring structure operations defined above, we have the following.

Power Series Raised to Powers

If n is a natural number we have

$$\left(\sum_{k=0}^{\infty} a_k X^k \right)^n = \sum_{m=0}^{\infty} c_m X^m,$$

where

$$c_0 = a_0^n, \quad c_m = \frac{1}{m a_0} \sum_{k=1}^{m} (kn - m + k) a_k c_{m-k},$$

for $m \geq 1$. (This formula can only be used if m and a_0 are invertible in the ring of scalars.)

In the case of formal power series with complex coefficients, the complex powers are well defined at least for series f with constant term equal to 1. In this case, f^α can be defined either by composition with the binomial series $(1+x)^\alpha$, or by composition with the exponential and the logarithmic series, $f^\alpha := \exp(\alpha \log(f))$, or as the solution of the differential equation $f(f^\alpha)' = \alpha f^\alpha f'$ with constant term 1, the three definitions being equivalent. The rules of calculus $(f^\alpha)^\beta = f^{\alpha\beta}$ and $f^\alpha g^\alpha = (fg)^\alpha$ easily follow.

Inverting Series

The series

$$A = \sum_{n=0}^{\infty} a_n X^n$$

in $\mathbb{R}[[X]]$ is invertible in $\mathbb{R}[[X]]$ if and only if its constant coefficient a_0 is invertible in R. This condition is necessary, for the following reason: if we suppose that A has an inverse $B = b_0 + b_1 x + \cdots$

then the constant term $a_0 b_0$ of $A \cdot B$ is the constant term of the identity series, i.e., it is 1. This condition is also sufficient; we may compute the coefficients of the inverse series B via the explicit recursive formula

$$b_0 = \frac{1}{a_0}$$

$$b_n = -\frac{1}{a_0} \sum_{i=1}^{n} a_i b_{n-i} \qquad \text{for } n \geq 1.$$

An important special case is that the geometric series formula is valid in $K[[X]]$:

$$(1-X)^{-1} = \sum_{n=0}^{\infty} X^n.$$

If $R = K$ is a field, then a series is invertible if and only if the constant term is non-zero, i.e., if and only if the series is not divisible by X. This says that $K[[X]]$ is a discrete valuation ring with uniformizing parameter X.

Dividing Series

The computation of a quotient $f / g = h$

$$\frac{\displaystyle\sum_{n=0}^{\infty} b_n X^n}{\displaystyle\sum_{n=0}^{\infty} a_n X^n} = \sum_{n=0}^{\infty} c_n X^n,$$

assuming the denominator is invertible (that is, a_0 is invertible in the ring of scalars), can be performed as a product f and the inverse of g, or directly equating the coefficients in $f = gh$:

$$c_n = \frac{1}{a_0} \left(b_n - \sum_{k=1}^{n} a_k c_{n-k} \right).$$

Extracting Coefficients

The coefficient extraction operator applied to a formal power series

$$f(X) = \sum_{n=0}^{\infty} a_n X^n$$

in X is written

$$\left[X^m \right] f(X)$$

and extracts the coefficient of X^m, so that

$$\left[X^m\right]f(X)=\left[X^m\right]\sum_{n=0}^{\infty}a_nX^n=a_m.$$

Composition of Series

Given formal power series

$$f(X)=\sum_{n=1}^{\infty}a_nX^n=a_1X+a_2X^2+\cdots$$

$$g(X)=\sum_{n=0}^{\infty}b_nX^n=b_0+b_1X+b_2X^2+\cdots,$$

one may form the *composition*

$$g(f(X))=\sum_{n=0}^{\infty}b_n(f(X))^n=\sum_{n=0}^{\infty}c_nX^n,$$

where the coefficients c_n are determined by "expanding out" the powers of $f(X)$:

$$c_n:=\sum_{k\in\mathbb{N},|j|=n}b_k a_{j_1}a_{j_2}\cdots a_{j_k}.$$

Here the sum is extended over all (k,j) with k in N and $j\in\mathbb{N}_+^k$ with $|j|:=j_1+\cdots+j_k=n$.

A more explicit description of these coefficients is provided by Faà di Bruno's formula, at least in the case where the coefficient ring is a field of characteristic 0.

A point here is that this operation is only valid when $f(X)$ has *no constant term*, so that each c_n depends on only a finite number of coefficients of $f(X)$ and $g(X)$. In other word the series for $g(f(X))$ converges in the topology of $R[[X]]$.

Example

Assume that the ring R has characteristic 0. If we denote by $\exp(X)$ the formal power series

$$\exp(X)=1+X+\frac{X^2}{2!}+\frac{X^3}{3!}+\frac{X^4}{4!}+\cdots,$$

then the expression

$$\exp(\exp(X)-1)=1+X+X^2+\frac{5X^3}{6}+\frac{5X^4}{8}+\cdots$$

makes perfect sense as a formal power series. However, the statement

$$\exp(\exp(X))=e\exp(\exp(X)-1)=e+eX+eX^2+\frac{5eX^3}{6}+\cdots$$

is not a valid application of the composition operation for formal power series. Rather, it is confusing the notions of convergence in $R[[X]]$ and convergence in R; indeed, the ring R may not even contain any number e with the appropriate properties.

Composition Inverse

Whenever a formal series $f(X) = \sum_k f_k X^k \in R[[X]]$ has $f_0 = 0$ and f_1 being an invertible element of R, there exists a series $g(X) = \sum_k g_k X^k$ that is the composition inverse of f, meaning that composing f with g gives the series representing the identity function (whose first coefficient is 1 and all other coefficients are zero). The coefficients of g may be found recursively by using the above formula for the coefficients of a composition, equating them with those of the composition identity X (that is 1 at degree 1 and 0 at every degree greater than 1). In the case when the coefficient ring is a field of characteristic 0, the Lagrange inversion formula provides a powerful tool to compute the coefficients of g, as well as the coefficients of the (multiplicative) powers of g.

Formal Differentiation of Series

Given a formal power series

$$f = \sum_{n \geq 0} a_n X^n$$

in $R[[X]]$, we define its formal derivative, denoted Df or f', by

$$Df = \sum_{n \geq 1} a_n n X^{n-1}.$$

The symbol D is called the formal differentiation operator. The motivation behind this definition is that it simply mimics term-by-term differentiation of a polynomial.

This operation is R-linear:

$$D(af + bg) = a \cdot Df + b \cdot Dg$$

for any a, b in R and any f, g in $R[[X]]$. Additionally, the formal derivative has many of the properties of the usual derivative of calculus. For example, the product rule is valid:

$$D(fg) = f \cdot (Dg) + (Df) \cdot g,$$

and the chain rule works as well:

$$D(f \circ g) = (Df \circ g) \cdot Dg,$$

whenever the appropriate compositions of series are defined.

Thus, in these respects formal power series behave like Taylor series. Indeed, for the f defined above, we find that

$$(D^k f)(0) = k!a_k,$$

where D^k denotes the kth formal derivative (that is, the result of formally differentiating k times).

Properties

Algebraic Properties of the Formal Power Series Ring

$R[[X]]$ is an associative algebra over R which contains the ring $R[X]$ of polynomials over R; the polynomials correspond to the sequences which end in zeros.

The Jacobson radical of $R[[X]]$ is the ideal generated by X and the Jacobson radical of R; this is implied by the element invertibility criterion discussed above.

The maximal ideals of $R[[X]]$ all arise from those in R in the following manner: an ideal M of $R[[X]]$ is maximal if and only if $M \cap R$ is a maximal ideal of R and M is generated as an ideal by X and $M \cap R$.

Several algebraic properties of R are inherited by $R[[X]]$:

- if R is a local ring, then so is $R[[X]]$;

- if R is Noetherian, then so is $R[[X]]$; this is a version of the Hilbert basis theorem;

- if R is an integral domain, then so is $R[[X]]$;

- if K is a field, then $K[[X]]$ is a discrete valuation ring.

Topological Properties of the Formal Power Series Ring

The metric space $(R[[X]], d)$ is complete.

The ring $R[[X]]$ is compact if and only if R is finite. This follows from Tychonoff's theorem and the characterisation of the topology on $R[[X]]$ as a product topology.

Applications

Formal power series can be used to solve recurrences occurring in number theory and combinatorics. For an example involving finding a closed form expression for the Fibonacci numbers.

One can use formal power series to prove several relations familiar from analysis in a purely algebraic setting. Consider for instance the following elements of $Q[[X]]$:

$$\sin(X) := \sum_{n \geq 0} \frac{(-1)^n}{(2n+1)!} X^{2n+1}$$

$$\cos(X) := \sum_{n \geq 0} \frac{(-1)^n}{(2n)!} X^{2n}$$

Then one can show that

$$\sin^2(X) + \cos^2(X) = 1,$$

$$\frac{\partial}{\partial X}\sin(X) = \cos(X),$$

$$\sin(X + Y) = \sin(X)\cos(Y) + \cos(X)\sin(Y).$$

The last one being valid in the ring Q[[X,Y]].

For K a field, the ring $K[[X_1, ..., X_r]]$ is often used as the "standard, most general" complete local ring over K in algebra.

Interpreting Formal Power Series as Functions

In mathematical analysis, every convergent power series defines a function with values in the real or complex numbers. Formal power series can also be interpreted as functions, but one has to be careful with the domain and codomain. If $f = \sum a_n X^n$ is an element of $R[[X]]$, S is a commutative associative algebra over R, I is an ideal in S such that the I-adic topology on S is complete, and x is an element of I, then we can define

$$f(X) = \sum_{n \geq 0} a_n X^n.$$

This latter series is guaranteed to converge in S given the above assumptions on X. Furthermore, we have

$$(f + g)(X) = f(X) + g(X)$$

and

$$(fg)(X) = f(X)g(X).$$

Unlike in the case of bona fide functions, these formulas are not definitions but have to be proved.

Since the topology on $R[[X]]$ is the (X)-adic topology and $R[[X]]$ is complete, we can in particular apply power series to other power series, provided that the arguments don't have constant coefficients (so that they belong to the ideal (X)): $f(0)$, $f(X^2-X)$ and $f((1-X)^{-1} - 1)$ are all well defined for any formal power series $f \in R[[X]]$.

With this formalism, we can give an explicit formula for the multiplicative inverse of a power series f whose constant coefficient $a = f(0)$ is invertible in R:

$$f^{-1} = \sum_{n \geq 0} a^{-n-1}(a - f)^n.$$

If the formal power series g with $g(0) = 0$ is given implicitly by the equation

$$f(g) = X$$

where f is a known power series with $f(0) = 0$, then the coefficients of g can be explicitly computed using the Lagrange inversion formula.

Generalizations

Formal Laurent Series

A formal Laurent series over a ring R is defined in a similar way to a formal power series, except that we also allow finitely many terms of negative degree (this is different from the classical Laurent series), that is series of the form

$$f = \sum_{n \in \mathbb{Z}} a_n X^n$$

where $a_n = 0$ for all but finitely many negative indices n. Multiplication of such series can be defined. Indeed, similarly to the definition for formal power series, the coefficient of X^k of two series with respective sequences of coefficients $\{a_n\}$ and $\{b_n\}$ is

$$\sum_{i \in \mathbb{Z}} a_i b_{k-i},$$

which sum is effectively finite because of the assumed vanishing of coefficients at sufficiently negative indices, and which sum zero for sufficiently negative k for the same reason.

For a non-zero formal Laurent series, the minimal integer n such that $a_n \neq 0$ is called the order of f, denoted ord (f). (The order of the zero series is $+\infty$.) The formal Laurent series form the ring of formal Laurent series over R, denoted by $R((X))$. It is equal to the localization of $R[[X]]$ with respect to the set of positive powers of X. It is a topological ring with the metric $d(f, g) = 2^{-\operatorname{ord}(f-g)}$.

If $R = K$ is a field, then $K((X))$ is in fact a field, which may alternatively be obtained as the field of fractions of the integral domain $K[[X]]$.

One may define formal differentiation for formal Laurent series in a natural way (term-by-term). Precisely, the formal derivative of the formal Laurent series f above is

$$f' = Df = \sum_{n \in \mathbb{Z}} n a_n X^{n-1}$$

which is again an element of $K((X))$. Notice that if f is a non-constant formal Laurent series, and K is a field of characteristic 0, then one has

$$\operatorname{ord}(f') = \operatorname{ord}(f) - 1.$$

However, in general this is not the case since the factor n for the lowest order term could be equal to 0 in R.

Formal Residue

Assume that K is a field of characteristic 0. Then the map

$$D : K((X)) \to K((X))$$

is a K-derivation that satisfies

$$\ker D = K$$

$$\operatorname{im} D = \{ f \in K((X)) : [X^{-1}] f = 0 \}.$$

The latter shows that the coefficient of X^{-1} in f is of particular interest; it is called *formal residue of f* and denoted $\operatorname{Res}(f)$. The map

$$\operatorname{Res} : K((X)) \to K$$

is K-linear, and by the above observation one has an exact sequence

$$0 \to K \to K((X)) \xrightarrow{D} K((X)) \xrightarrow{\operatorname{Res}} K \to 0.$$

Some rules of calculus. As a quite direct consequence of the above definition, and of the rules of formal derivation, one has, for any $f, g \in K((X))$

    i. $\operatorname{Res}(f') = 0$;

    ii. $\operatorname{Res}(fg') = -\operatorname{Res}(f'g)$;

    iii. $\operatorname{Res}(f' / f) = \operatorname{ord}(f), \qquad \forall f \neq 0$;

    iv. $\operatorname{Res}\big((f \circ g) g'\big) = \operatorname{ord}(g) \operatorname{Res}(f)$, if $\operatorname{ord}(g) > 0$;

    v. $[X^n] f(X) = \operatorname{Res}\big(X^{-n-1} f(X)\big)$.

Property (i) is part of the exact sequence above. Property (ii) follows from (i) as applied to $(fg)' = f'g + fg'$. Property (iii): any f can be written in the form $f = x^m g$, with $m = \operatorname{ord}(f)$. and $\operatorname{ord}(g) = 0$: then $f' / f = mX^{-1} + g' / g$. Since $\operatorname{ord}(g) = 0$, the element g is invertible in $K[[X]] \subset \operatorname{im}(D) = \ker(\operatorname{Res})$, whence $\operatorname{Res}(f' / f) = m$. Property (iv): Since $\operatorname{im}(D) = \ker(\operatorname{Res})$, we can write $f = f_{-1} X^{-1} + F'$, with $F \in K[[X]]$. Consequently, $(f \circ g) g' = f_{-1} g^{-1} g' + (F' \circ g) g' = f_{-1} g' / g + (F \circ g)'$ and (iv) follows from (i) and (iii). Property (v) is clear from the definition.

The Lagrange Inversion Formula

As mentioned above, any formal series $f \in K[[X]]$ with $f_0 = 0$ and $f_1 \neq 0$ has a composition inverse g in $K[[X]]$. The following relation between the coefficients of g^n and f^{-k} holds ("Lagrange inversion formula"):

$$k[X^k] g^n = n[X^{-n}] f^{-k}.$$

In particular, for $n = 1$ and all $k \geq 1$,

$$[X^k]g = \frac{1}{k}\text{Res}\left(f^{-k}\right).$$

Since the proof of the Lagrange inversion formula is a very short computation, it is worth reporting it here. Since $\text{ord}(f) = 1$, by the above rules of calculus,

$$k\left[X^k\right]g^n \quad \aleph k\text{Res}\left(g^n X^{-k-\ddot{u}}\right) \quad k\text{ord}(f)\text{Res}\left(g^n X^{-k-}\right) \quad k\text{Res}\left(((g^n X^{-k-}) \circ f)f'\right)$$

$$= k\text{Res}\left((g \circ f)^n f^{-k-1} f'\right) = k\text{Res}\left(X^n f^{-k-1} f'\right) = -\text{Res}\left(X^n (f^{-k})'\right)$$

$$= \text{Res}\left((X^n)' f^{-k}\right) = n\text{Res}\left(X^{n-1} f^{-k}\right) = n[X^{-n}]f^{-k}.$$

Generalizations. One may observe that the above computation can be repeated plainly in more general settings than $K((X))$: a generalization of the Lagrange inversion formula is already available working in the $C((X))$-modules $X^\alpha C((X))$, where α is a complex exponent. As a consequence, if f and g are as above, with $f_1 = g_1 = 1$, we can relate the complex powers of f/X and g/X: precisely, if α and β are non-zero complex numbers with negative integer sum, $m = -\alpha - \beta \in \mathbb{N}$, then

$$\frac{1}{\alpha}[X^m]\left(\frac{f}{X}\right)^\alpha = -\frac{1}{\beta}[X^m]\left(\frac{g}{X}\right)^\beta.$$

For instance, this way one finds the power series for complex powers of the Lambert function.

Power Series in Several Variables

Formal power series in any number of indeterminates (even infinitely many) can be defined. If I is an index set and X_I is the set of indeterminates X_i for $i \in I$, then a monomial X^α is any finite product of elements of X_I (repetitions allowed); a formal power series in X_I with coefficients in a ring R is determined by any mapping from the set of monomials X^α to a corresponding coefficient c_α, and is denoted $\sum_\alpha c_\alpha X^\alpha$. The set of all such formal power series is denoted $R[[X_I]]$, and it is given a ring structure by defining

$$\left(\sum_\alpha c_\alpha X^\alpha\right) + \left(\sum_\alpha d_\alpha X^\alpha\right) = \sum_\alpha (c_\alpha + d_\alpha)X^\alpha$$

and

$$\left(\sum_\alpha c_\alpha X^\alpha\right) \times \left(\sum_\beta d_\beta X^\beta\right) = \sum_{\alpha,\beta} c_\alpha d_\beta X^{\alpha+\beta}$$

Topology

The topology on $R[[X_I]]$ is such that a sequence of its elements converges only if for each monomial X^α the corresponding coefficient stabilizes. If I is finite, then this the J-adic topology, where J is the ideal of $R[[X_I]]$ generated by all the indeterminates in X_I. This does not hold if I is infinite. For example, if $I = N$, then the sequence $(f_n)_{n \in N}$ with $f_n = X_n + X_{n+1} + X_{n+2} + \dots$ does not converge with respect to any J-adic topology on R, but clearly for each monomial the corresponding coefficient stabilizes.

As remarked above, the topology on a repeated formal power series ring like $R[[X]][[Y]]$ is usually chosen in such a way that it becomes isomorphic as a topological ring to $R[[X,Y]]$.

Operations

All of the operations defined for series in one variable may be extended to the several variables case.

- A series is invertible if and only if its constant term is invertible in R.

- The composition $f(g(X))$ of two series f and g is defined if f is a series in a single indeterminate, and the constant term of g is zero. For a series f in several indeterminates a form of "composition" can similarly be defined, with as many separate series in the place of g as there are indeterminates.

In the case of the formal derivative, there are now separate partial derivative operators, which differentiate with respect to each of the indeterminates. They all commute with each other.

Universal Property

In the several variables case, the universal property characterizing $R[[X_1, \dots, X_r]]$ becomes the following. If S is a commutative associative algebra over R, if I is an ideal of S such that the I-adic topology on S is complete, and if $x_1, \dots, x_r$ are elements of I, then there is a *unique* $\Phi : R[[X_1, \dots, X_n]] \to S$ with the following properties:

- Φ is an R-algebra homomorphism

- Φ is continuous

- $\Phi(X_i) = x_i$ for $i = 1, \dots, r$.

Non-commuting Variables

The several variable case can be further generalised by taking *non-commuting variables* X_i for $i \in I$, where I is an index set and then a monomial X^α is any word in the X_I; a formal power series in X_I with coefficients in a ring R is determined by any mapping from the set of monomials X^α to a corresponding coefficient c_α, and is denoted $\sum_\alpha c_\alpha X^\alpha$. The set of all such formal power series is denoted $R«X_I»$, and it is given a ring structure by defining addition pointwise

$$\left(\sum_\alpha c_\alpha X^\alpha \right) + \left(\sum_\alpha d_\alpha X^\alpha \right) = \sum_\alpha (c_\alpha + d_\alpha) X^\alpha$$

and multiplication by

$$\left(\sum_\alpha c_\alpha X^\alpha\right)\times\left(\sum_\alpha d_\alpha X^\alpha\right)=\sum_{\alpha,\beta}c_\alpha d_\beta X^\alpha\cdot X^\beta$$

where $\cdot$ denotes concatenation of words. These formal power series over R form the Magnus ring over R.

On a Semiring

In theoretical computer science, the following definition of a formal power series is given: let Σ be an alphabet (finite set) and S be a semiring. In this context, a formal power series is any mapping r from the set of strings generated by Σ (denoted as Σ^*) to the semiring S. The values of such a mapping r are (somewhat idiosyncratically) denoted as (r, w) were $w \in \Sigma^*$. Then the mapping r itself is conventionally written as $r = \sum_{w\in\Sigma^*} (r, w)w$. Given this notation, the values (r, w) are also called the coefficients of the series. Similarly to the non-commuting [ring] case discussed in the section above this, the notation for the collection of all power series given a fixed alphabet and semiring is $S \langle\!\langle \Sigma^* \rangle\!\rangle$.

Replacing the Index set by an Ordered Abelian Group

Suppose G is an ordered abelian group, meaning an abelian group with a total ordering respecting the group's addition, so that $a < b$ if and only if $a + c < b + c$ for all c. Let I be a well-ordered subset of G, meaning I contains no infinite descending chain. Consider the set consisting of

$$\sum_{i\in I} a_i X^i$$

for all such I, with a_i in a commutative ring R, where we assume that for any index set, if all of the a_i are zero then the sum is zero. Then $R((G))$ is the ring of formal power series on G; because of the condition that the indexing set be well-ordered the product is well-defined, and we of course assume that two elements which differ by zero are the same. Sometimes the notation $[[R^G]]$ is used to denote $R((G))$.

Various properties of R transfer to $R((G))$. If R is a field, then so is $R((G))$. If R is an ordered field, we can order $R((G))$ by setting any element to have the same sign as its leading coefficient, defined as the least element of the index set I associated to a non-zero coefficient. Finally if G is a divisible group and R is a real closed field, then $R((G))$ is a real closed field, and if R is algebraically closed, then so is $R((G))$.

This theory is due to Hans Hahn, who also showed that one obtains subfields when the number of (non-zero) terms is bounded by some fixed infinite cardinality.

Examples and Related Topics

- Bell series are used to study the properties of multiplicative arithmetic functions

- Formal groups are used to define an abstract group law using formal power series

- Puiseux series are an extension of formal Laurent series, allowing fractional exponents

- Rational series

We will first try to develop the theory of generating functions by getting closed form expressions for some known recurrence relations. These ideas will be used later to get some binomial identities. $n \in \mathbb{Q}$ and $k \in \mathbb{Z}$, $k \geq 0$, the binomial coefficients, $\binom{n}{k}$, are well defined, using the idea that $\binom{n}{k} = \dfrac{n(n-1)(n-2)\cdots(n-k+1)}{k!}$. We now start with the definition of "formal power series" over $\mathbb{Q}$ and study its properties in some detail.

Definition: (Formal power series). An algebraic expression of the form $f(x) = \sum_{n \geq 0} a_n x^n$, where $a_n \in \mathbb{Q}$ for all n ≥ 0, is called a formal power series in the indeterminate x over $\mathbb{Q}$.

The set of all formal power series in the indeterminate x, with coefficients from $\mathbb{Q}$ will be denoted by $\mathcal{P}(x)$.

Remark: Given a sequence of numbers $\{a_n \in \mathbb{Q} : n = 0, 1, 2, \ldots\}$, one associates two formal power series, namely, $\sum_{n \geq 0} a_n x^n$ and $\sum_{n \geq 0} a_n \dfrac{x^n}{n!}$. The expression $\sum_{n \geq 0} a_n x^n$ is called the generating function and the expression $\sum_{n \geq 0} a_n \dfrac{x^n}{n!}$ is called the exponential generating function, for the numbers $\{a_n : n \geq 0\}$.

2. Let $f(x) = \sum_{n \geq 0} a_n x^n$ be a formal power series. Then the coefficient of x^n, for n ≥ 0, in $f(x)$ is denoted by $\left[x^n\right]f(x)$. That is, $a_0 = \left[x^0\right]f(x)$ and $a_n = \left[x^n\right]f(x)$, for n ≥ 1.

3. One thinks of $\sum_{n \geq 0} a_n x^n$ as an algebraic expression. In general, one is interested only in computing the coefficient of certain power of x and not in evaluating them for any value of x. But, if at all there is a need to evaluate it at a point, say x_0, then one needs to determine its "radius of convergence" and then evaluate it if x_0 lies within that radius.

We need the following definition to proceed further.

Definition: (Equality of two formal power series). Two elements $f(x) = \sum_{n \geq 0} a_n x^n$ and $g(x) = \sum_{n \geq 0} b_n x^n$ of $\mathcal{P}(x)$ are said to be equal if $a_n = b_n$, for all n ≥ 0.

We are now ready to define the algebraic rules:

Definition: Let $f(x) = \sum_{n \geq 0} a_n x^n, g(x) = \sum_{n \geq 0} b_n x^n \in \mathcal{P}(x)$. Then their

1. sum/addition is defined by

$$f(x) + g(x) = \sum_{n \geq 0} a_n x^n + \sum_{n \geq 0} b_n x^n = \sum_{n \geq 0} (a_n + b_n)x^n.$$

2. product is defined by

$$f(x) \cdot g(x) = \left(\sum_{n \geq 0} a_n x^n \right) \cdot \left(\sum_{n \geq 0} b_n x^n \right) = \sum_{n \geq 0} c_n x^n, \ where \ c_n = \sum_{k=0}^{n} a_k b_{n-k}, \ for \ n \geq 0.$$

This product is also called the Cauchy product.

Remark: In case of exponential power series, the product of $f(x) = \sum_{n \geq 0} a_n \dfrac{x^n}{n!}$ and $g(x) = \sum_{n \geq 0} b_n \dfrac{x^n}{n!}$ equals $\sum_{n \geq 0} d_n \dfrac{x^n}{n!}$, where $d_n = \sum_{k=0}^{n} \binom{n}{k} a_k b_{n-k} \ for \ n \geq 0$

2. Note that the expression $e^{e^x - 1}$ is a well defined formal power series as the definition $e^y = \sum_{n \geq 0} \dfrac{y^n}{n!}$ implies that $e^{e^x - 1} = \sum_{n \geq 0} \dfrac{(e^x - 1)^n}{n!}$ and hence

$$\left[x^m \right] e^{e^x - 1} = \left[x^m \right] \sum_{n \geq 0} \dfrac{(e^x - 1)^n}{n!} = \sum_{n=0}^{m} \left[x^m \right] \dfrac{(e^x - 1)^n}{n!}.$$

That is, for each $m \geq 0, \left[x^m \right] e^{e^x - 1}$ is a sum of a finite number of real numbers. Where as the expression e^{e^x} is not a formal power series as the computation of $\left[x^m \right] e^{e^x}$, for all m ≥ 0, will indeed require an infinite sum.

Thus, under the algebraic operations defined above, it can be checked that the set $\mathcal{P}(x)$ forms a Commutative Ring with identity, where the identity element is given by the formal power series $f(x) = 1$. In this ring, the element $f(x) = \sum_{n \geq 0} a_n x^n$ is said to have a reciprocal if there exists an other element $g(x) = \sum_{n \geq 0} b_n x^n \in P(x)$ such that $f(x) \cdot g(x) = 1$. So, the question arises, under what conditions on the coefficients of $f(x)$, can we find $g(x) \in \mathcal{P}(x)$ such that $f(x)g(x) = 1$.

The answer to this question is given in the following proposition.

Proposition: Let $f(x) = \sum_{n \geq 0} a_n x^n \in \mathcal{P}(x)$. Then there exists $f(x) \in \mathcal{P}(x)$ satisfying $f(x) \cdot g(x) = 1$ if and only if $a_0 \neq 0$.

Proof. Let $g(x) = \sum_{n \geq 0} b_n x^n \in P(x)$. Then, by the definition of Cauchy product, $f(x)g(x) = \sum_{n \geq 0} c_n x^n$ where $c_n = \sum_{n \geq 0}^{n} a_k b_{n-k}$, for all n ≥ 0. Therefore, using the definition of equality of two power series, we see that $f(x)g(x) = 1$ if and only if $c_0 = 1$ and $c_n = 0$, for all n ≥ 1.

Therefore, if $a_0 = 0$ then $c_0 = 0$ and hence the Cauchy product $f(x)g(x)$ can never equal 1.

However, if $a_0 \neq 0$, then the coefficients b_n's can be recursively obtained as follows:

$$b_0 = \frac{1}{a_0} \ as \ 1 = c_0 = a_0 b_0$$

$$b_1 = \frac{-1}{a_0} \cdot (a_1 b_0) \ as \ 0 = c_1 = a_0 b_1 + a_1 b_0.$$

$b_2 = \frac{-1}{a_0} \cdot (a_2 b_0 + a_1 b_1) \ as \ 0 = c_2 = a_0 b_2 + a_1 b_1 + a_2 b_0.$ And in general, if we have already computed the values of b_k, for $k \le r$, then

$$b_{r+1} = \frac{-1}{a_0} \cdot (a_{r+1} b_0 + a_r b_1 + \cdots a_1 b_r) \ as \ 0 = c_{r+1} = a_{r+1} b_0 + a_r b_1 + \cdots + a_1 b_r + a_0 b_{r+1}.$$

Note that if the coefficients a_n's come from $\mathcal{R}$, a commutative ring with unity, then one needs bo to be an invertible element of $\mathcal{R}$ for Proposition 2.4.6 to hold true. Let us now look at the composition of two formal power series. Recall that, if $f(x) = \sum_{n \ge 0} a_n x^n$, $g(x) = \sum_{n \ge 0} b_n x^n \in \mathcal{P}(x)$ then the composition $((f \circ g)(x) = f(g(x)) = \sum_{n \ge 0} a_n (g(x))^n = \sum_{n \ge 0} a_n (\sum_{m \ge 0} b_m x^m)^n$ may not be defined (just to compute the constant term of the composition, one may have to look at an infinite sum). For example, let $f(x) = e^x$ and $g(x) = x + 1$. Note that $g(0) = 1 \ne 0$. Here, $(f \circ g)(x) = f(g(x)) = f(x+1) = e^{x+1}$. So, as function $f \circ g$ is well defined, but there is no formal procedure to write e^{x+1} as $\sum_{k \ge 0} a_k x^k \in \mathcal{P}(x)$ (i.e., with $a_k \in \mathbb{Q}$) and hence e^{x+1} is not a formal power series over $\mathbb{Q}$.

The next result gives the condition under which the composition $(f \circ g)(x)$ is well defined.

Proposition: Let $f(x) = \sum_{n \ge 0} a_n x^n$ and $g(x) = \sum_{n \ge 0} b_n x^n$ be two formal power series. Then the composition $(f \circ g)(x)$ is well defined if either f is a polynomial or $b_0 = 0$.

Moreover, suppose that $a_0 = 0$. Then, there exists $g(x) = \sum_{n \ge 0} b_n x^n$, with $b_0 = 0$, such that $(f \circ g)(x) = x$. Furthermore, $(g \circ f)(x)$ is well defined and $(g \circ f)(x) = x$.

Proof. Let $(f \circ g)(x) = f(g(x)) = \sum_{n \ge 0}^{n} c_n x^n$ and suppose that either f is a polynomial or $b_0 = 0$. Then to compute $c_k = [x^k](f \circ g)(x)$, for k ≥ 0, one just needs to consider the terms $a_0 + a_1 g(x) + a_2 (g(x))^2 + \cdots + a_k (g(x))^k$. Hence, each c_k is a real number and $(f \circ g)(x)$ is well defined. This completes the proof of the first portion. The proof of the other part is left to the readers.

We now define the formal differentiation of elements of $\mathcal{P}(x)$ and state a result without proof.

Definition: (Differentiation). Let $f(x) = \sum_{n \ge 0} a_n x^n \in \mathcal{P}(x)$. Then the formal differentia-tion of $f(x)$, denoted $Df(x)$, is defined by

$$Df(x) = a_1 + 2a_2 x + \cdots + na_n x^{n-1} + \cdots = \sum_{n \ge 0} na_n x^{n-1}.$$

Proposition: Let $f(x) = \sum_{n \geq 0} a_n x^n \in P(x)$. Then $f(x) = a_0$, a constant, whenever $Df(x) = 0$. Also, $f(x) = a_0 e^x$ whenever $Df(x) = f(x)$.

Before proceeding further, let us look at some important examples, where the above results can be used.

Example: Using the generalized binomial theorem, recall that, for a positive integer r, the formal power series

$$\sum_{n \geq 0} \binom{n+r-1}{n} x^n \ equals \ \frac{1}{(1-x)^r}.$$

In particular, for r = 1, the formal power series $\sum_{n \geq 0} nx^n$ equals $\frac{1}{(1-x)}$.

1. Determine a closed form expression for $\sum_{n \geq 0} nx^n \in P(x)$.

Solution: As $\frac{1}{1-x} = \sum_{n \geq 0} x^n$, one has $\frac{1}{(1-x)^2} = D\left(\frac{1}{1-x}\right) = D\left(\sum_{n \geq 0} x^n\right) = \sum_{n \geq 0} nx^{n-1}$.

Thus, the closed form expression is $\frac{1}{(1-x)^2}$.

This can also be computed as follows:

Let $S = \sum_{n \geq 0} nx^n = x + 2x^2 + 3x^3 + \cdots$. Then $xS = x^2 + 2x^3 + 3x^4 + \cdots$. Hence,

$$(1-x)S = x + x^2 + x^3 + \cdots = x(1 + x + x^2 + \cdots) = \frac{x}{1-x}. \ Thus, \ S = \frac{1}{(1-x)^2}.$$

2. Let $f(x) = \sum_{n \geq 0} a_n x^n \in P(x)$. Determine $\sum_{k \geq 0}^{n} a_k$.

Solution: Recall that the Cauchy product of $f(x) = \sum_{n \geq 0} a_n x^n$ and $g(x) = \sum_{n \geq 0} b_n x^n$ equals $\sum_{n \geq 0} c_n x^n$ where $c_n = \sum_{k=0}^{n} a_k b_{n-k}$, for n ≥ 0. Therefore, to get $c_n = \sum_{k=0}^{n} a_k$, one needs b_k = 1, for all k ≥ 0. That is, $\sum_{k=0}^{n} a_k = c_n [x^n]\left(f(x) \cdot \frac{1}{1-x}\right)$.

Hence, the Cauchy product helps us in computing the sum of the first N coefficients of a formal power series, for any N ≥ 1.

3. Determine the sum of the squares of the first N positive integers.

Solution: Using Equation (1), observe that $k = [x^{k-1}]\left(\frac{1}{(1-x)^2}\right)$. Therefore, using example, one has

$$\sum_{k=1}^{N} k = \left[x^{N-1}\right]\left(\frac{1}{(1-x)^2} \cdot \frac{1}{1-x}\right) = \left[x^{N-1}\right]\frac{1}{(1-x)^3} = \binom{N-1+3-1}{N-1} = \frac{N(N+1)}{2}.$$

4. Determine a closed form expression for $\sum_{k=1}^{N} k^2$.

Solution: Using example, observe that $\sum_{k\geq 0} kx^k = \frac{1}{(1-x)^2}$. Therefore, using the differentiation operator, one obtains

$$\sum_{k\geq 0} k^3 x^k = x\left(\sum_{k\geq 0} k^2 x^{k-1}\right) = xD\left(\frac{x}{(1-x)^2}\right) = \frac{x(1+x)}{(1-x)^3} .$$

Thus, by Example

$$\sum_{k=1}^{N} k^2 = \left[x^N\right]\left(\frac{x(1+x)}{(1-x)^3} \cdot \frac{1}{1-x}\right) = \left[x^{N-1}\right]\left(\frac{1}{(1-x)^4}\right) + \left[x^{N-2}\right]\left(\frac{1}{(1-x)^4}\right)$$

$$= \binom{N-1+4-1}{N-1} + \binom{N-2+4-1}{N-2}$$

$$= \frac{N(N+1)(2N+1)}{6}.$$

5. Determine a closed form expression for $\sum_{k=1}^{N} k^3$.

Solution: Using Equation (2), observe that $\sum_{k\geq 0} k^2 x^k = \frac{x(1+x)}{(1-x)^3}$. So,

$$\sum_{k\geq 0} k^2 x^k = x\left(\sum_{k\geq 0} k^2 x^{k-1}\right) = xD\left(\frac{x(1+x)}{(1-x)^3}\right) = \frac{x(1+4x+x^2)}{(1-x)^4}.$$

Thus, by Example

$$\sum_{k=1}^{N} k^3 = \left[x^N\right]\left(\frac{x(1+4x+x^2)}{(1-x)^4)} \cdot \frac{1}{1-x}\right)$$

$$= \left[x^{N-1}\right]\left(\frac{1}{(1-x)^5}\right) + \left[x^{N-2}\right]\left(\frac{4}{(1-x)^5}\right) + \left[x^{N-3}\right]\left(\frac{1}{(1-x)^5}\right)$$

$$= \binom{N-1+5-1}{N-1} + 4\binom{N-2+5-1}{N-2} + \binom{N-3+5-1}{N-3}$$

$$= \left(\frac{N(N+1)}{2}\right)^2.$$

Hence, we observe that we can inductively use this technique to get a closed form expression for $\sum_{k=1}^{N} k^r$, for any positive integer r.

6. Determine a closed form expression for $\sum_{n \geq 0} \dfrac{n^2 + n + 6}{n!}$.

Solution: As we need to compute the infinite sum, Cauchy product cannot be used. Also, one needs to find a convergent series, which when evaluated at some x_0, gives the required expression. Therefore, recall that the series $e^x = \sum_{n \geq 0} \dfrac{x^n}{n!}$, converges for all $x \in \mathbb{R}$ and evaluating e^x at $x = 1$, gives $\sum_{n \geq 0} \dfrac{1}{n!} = e$.

Similarly, $\dfrac{n^2}{n!} = \left[x^n\right]\left(xD\left(xDe^x\right)\right) = \left[x^n\right]\left((x + x^2)e^x\right)$ and $\dfrac{n}{n!} = \left[x^n\right]\left(xD\left(xDe^x\right)\right) = \left[x^n\right]\left((x + x^2)e^x\right)$. Thus,

$$\sum_{n \geq 0} = \dfrac{n^2 + n + 6}{n!} = (x + x^2)e^x + xe^x + 6e^x \big|_{x=1} = 9e.$$

7. Let n and r be two fixed positive integers. Then determine the number of non-negative integer solutions to the system $x_1 + x_2 + \cdots + x_n = r$.

Solution: Recall that this number equals $\begin{pmatrix} r + n - 1 \\ r \end{pmatrix}$

We can think of the problem as follows: the above system can be interpreted as coming from the monomial x^r, where $r = x_1 + x_2 + \cdots + x_n$. That is, the problem reduces to finding the coefficients of y^{x_k} of a formal power series, for non-negative integers x_k's.

Now, recall that the terms y^{x_k} appear with a coefficient 1 in the expression $\dfrac{1}{1 - y} = \sum_{i \geq 0} y^i$.

Hence, the question reduces to computing

$$\left[y^r\right] = \left(\dfrac{1}{(1 - y)(1 - y)\cdots(1 - y)}\right) = \left[y^r\right]\dfrac{1}{(1 - y)^n} = \begin{pmatrix} r + n - 1 \\ r \end{pmatrix}.$$

We now look at some examples that may require the use of the package "MATHEMATICA" or "MAPLE" or ... to obtain the exact answer. So, in the examples given below, one is interested in getting only a formal power series whose coefficient gives the required result even though it may be difficult to compute the coefficient.

Example: For fixed positive integers n and r, determine the number of non-negative integer solutions to the system $x_1 + 2x_2 + \cdots + nx_n = r$.

Solution: The ideas from example imply that one needs to consider the formal power series $\sum_{i \geq 0} x^{ki} = \dfrac{1}{1 - x^k}$. Hence, need to compute

$$\left[x^r\right]\dfrac{1}{(1 - x)(1 - x^2)\cdots(1 - x^n)}.$$

2. Determine the number of solutions to the system $x_1 + x_2 + \cdots + x_5 = n$, where x_i's are non-negative integer, $x_1 \geq 4$, $x_4 \leq 10$ and x_r is a multiple of r, whenever $r \neq 1, 4$.

Solution: The condition $x_1 \geq 4$ corresponds to looking at x^k, for $k \geq 4$, which corresponds to the formal power series $\sum_{k \geq 4} x^k$. Similarly, $x_4 \leq 10$ gives the formal power series $\sum_{k=0}^{10} x^k$ and the condition x_r is a multiple of r, for $r \neq 1, 4$, corresponds to $\sum_{k \geq 0} x^{rk}$. So, we need to compute the coefficient of x^n in the product

$$\left(\sum_{k \geq 4} x^k\right) \cdot \left(\sum_{k=0}^{10} x^k\right) \cdot \left(\sum_{k \geq 0} x^{2k}\right) \cdot \left(\sum_{k \geq 0} x^{3k}\right) \cdot \left(\sum_{k \geq 0} x^{5k}\right) = \frac{x^4(1-x^{11})}{(1-x)^2(1-x^2)(1-x^3)(1-x^5)}.$$

3. Determine the number of ways in which 100 voters can cast their 100 votes for 10 candidates such that no candidate gets more than 20 votes.

Solution: Note that we are assuming that the voters are identical. So, we need to solve the system in non-negative integers to the system $x_1 + x_2 + \cdots + x_{10} = 100$, with $0 \leq x_i \leq 20$, for $1 \leq i \leq 10$. So, we need to find the coefficient of x^{100} in

$$\left(\sum_{k=1}^{20} x^k\right)^{10} = \frac{(1-x^{21})^{10}}{(1-x)^{10}} = \left(\sum_{i=0}^{10} (-1)^i \binom{10}{i} x^{21i}\right) \cdot \left(\sum_{j \geq 0} \binom{10+j-1}{j} x^j\right).$$

Thus, the required answer equals $[x^{100}] \left(\sum_{k=1}^{20} x^k\right)^{10} = \sum_{i=0}^{4} (-1)^i \binom{10}{i} \cdot \binom{109-21i}{9}.$

Before moving to the applications of formal power series to the solution of recurrence rela- tions, let us list a few well known power series. The readers are requested to get a proof for their satis- faction.

Table of Formal Power Series

$$e^x = \sum_{k \geq 0} \frac{x^k}{k!} \qquad\qquad \log(1-x) = -\sum_{k \geq 1} \frac{x^k}{k}, |x| < 1$$

$$(1+x)^a = \sum_{r \geq 0} \binom{a}{k} x^k, |x| < 1 \qquad\qquad \frac{1}{1-x} = \sum_{k \geq 0} x^k, |x| < 1$$

$$\frac{1}{(1-x)^a} = \sum_{k \geq 0} \binom{a+k-1}{k} x^k, |x| < 1 \qquad\qquad \frac{1}{\sqrt{1-4x}} = \sum_{k \geq 0} \binom{2k}{k} x^k, |x| < \frac{1}{4}$$

$$x^{-r}(1+x)^n = \sum_{k \geq -r} \binom{n}{r+k} x^k, |x| < 1 \qquad\qquad \frac{x^n}{(1-x)^{n+1}} = \sum_{k \geq 0} \binom{k}{n} x^k, n \geq 0, |x| < 1$$

$$\sin(x) = \sum_{r \geq 0} \frac{(-1)^r x^{2r+1}}{(2r+1)!} \qquad\qquad \cos(x) = \sum_{r \geq 0} \frac{(-1)^r x^{2r}}{(2r)!}$$

$$\sinh(x) = \sum_{r \geq 0} \frac{x^{2r+1}}{(2r+1)!} \qquad\qquad \cosh(x) = \sum_{r \geq 0} \frac{x^{2r}}{(2r)!}$$

$$\frac{1-\sqrt{1-4x}}{2x} = \sum_{k \geq 0} \frac{1}{k+1} \binom{2k}{k} x^k, |x| < \frac{1}{4}$$

Recurrence Relation

In mathematics, a recurrence relation is an equation that recursively defines a sequence or multidimensional array of values, once one or more initial terms are given: each further term of the sequence or array is defined as a function of the preceding terms.

The term difference equation sometimes refers to a specific type of recurrence relation. However, "difference equation" is frequently used to refer to *any* recurrence relation.

Examples

Logistic Map

An example of a recurrence relation is the logistic map:

$$x_{n+1} = rx_n(1-x_n),$$

with a given constant r; given the initial term x_0 each subsequent term is determined by this relation.

Some simply defined recurrence relations can have very complex (chaotic) behaviours, and they are a part of the field of mathematics known as nonlinear analysis.

Solving a recurrence relation means obtaining a closed-form solution: a non-recursive function of n.

Fibonacci Numbers

The recurrence satisfied by the Fibonacci numbers is the archetype of a homogeneous linear recurrence relation with constant coefficients. The Fibonacci sequence is defined using the recurrence

$$F_n = F_{n-1} + F_{n-2}$$

with seed values

$$F_0 = 0$$
$$F_1 = 1.$$

Explicitly, the recurrence yields the equations

$$F_2 = F_1 + F_0$$
$$F_3 = F_2 + F_1$$
$$F_4 = F_3 + F_2 \text{ etc.}$$

We obtain the sequence of Fibonacci numbers, which begins

$$0, 1, 1, 2, 3, 5, 8, 13, 21, 34, 55, 89, \ldots$$

The recurrence can be solved by methods described below yielding Binet's formula, which involves powers of the two roots of the characteristic polynomial $t^2 = t + 1$; the generating function of the sequence is the rational function

$$\frac{t}{1-t-t^2}.$$

Binomial Coefficients

A simple example of a multidimensional recurrence relation is given by the binomial coefficients $\binom{n}{k}$, which count the number of ways of selecting k elements out of a set of n elements. They can be computed by the recurrence relation

$$\binom{n}{k} = \binom{n-1}{k-1} + \binom{n-1}{k},$$

with the base cases $\binom{n}{0} = \binom{n}{n} = 1$. Using this formula to compute the values of all binomial coefficients generates an infinite array called Pascal's triangle. The same values can also be computed directly by a different formula that is not a recurrence, but that requires multiplication and not just addition to compute:

$$\binom{n}{k} = \frac{n!}{k!(n-k)!}.$$

Relationship to Difference Equations Narrowly Defined

Given an ordered sequence $\{a_n\}_{n=1}^{\infty}$ of real numbers: the first difference $\Delta(a_n)$ is defined as

$$\Delta(a_n) = a_{n+1} - a_n$$

The second difference $\Delta^2(a_n)$ is defined as

$$\Delta^2(a_n) = \Delta(a_{n+1}) - \Delta(a_n),$$

which can be simplified to

$$\Delta^2(a_n) = a_{n+2} - 2a_{n+1} + a_n$$

More generally: the kth difference of the sequence a_n written as $\Delta^k(a_n)$ is defined recursively as

$$\Delta^k(a_n) = \Delta^{k-1}(a_{n+1}) - \Delta^{k-1}(a_n) = \sum_{t=0}^{k} \binom{k}{t}(-1)^t a_{n+k-t}$$

(The sequence and its differences are related by a binomial transform.) The more restrictive definition of difference equation is an equation composed of a_n and its kth differences. (A widely used broader definition treats "difference equation" as synonymous with "recurrence relation".)

Actually, it is easily seen that,

$$a_{n+k} = \binom{k}{0}a_n + \binom{k}{1}\Delta(a_n) + \cdots + \binom{k}{k}\Delta^k(a_n)$$

Thus, a difference equation can be defined as an equation that involves a_n, a_{n-1}, a_{n-2} etc. (or equivalenty a_n, a_{n+1}, a_{n+2} etc.)

Since difference equations are a very common form of recurrence, some authors use the two terms interchangeably. For example, the difference equation

$$3\Delta^2(a_n) + 2\Delta(a_n) + 7a_n = 0$$

is equivalent to the recurrence relation

$$3a_{n+2} = 4a_{n+1} - 8a_n$$

Thus one can solve many recurrence relations by rephrasing them as difference equations, and then solving the difference equation, analogously to how one solves ordinary differential equations. However, the Ackermann numbers are an example of a recurrence relation that do not map to a difference equation, much less points on the solution to a differential equation.

Summation equations relate to difference equations as integral equations relate to differential equations.

From Sequences to Grids

Single-variable or one-dimensional recurrence relations are about sequences (i.e. functions defined on one-dimensional grids). Multi-variable or n-dimensional recurrence relations are about n-dimensional grids. Functions defined on n-grids can also be studied with partial difference equations.

Solving

Roots of the Characteristic Polynomial

An order-d homogeneous linear recurrence with constant coefficients is an equation of the form

$$a_n = c_1 a_{n-1} + c_2 a_{n-2} + \cdots + c_d a_{n-d},$$

where the d coefficients c_i (for all i) are constants.

A constant-recursive sequence is a sequence satisfying a recurrence of this form. There are d degrees of freedom for solutions to this recurrence, i.e., the initial values $a_0, \ldots, a_{d-1}$ can be taken to be any values but then the recurrence determines the sequence uniquely.

The same coefficients yield the characteristic polynomial (also "auxiliary polynomial")

$$p(t) = t^d - c_1 t^{d-1} - c_2 t^{d-2} - \cdots - c_d$$

whose d roots play a crucial role in finding and understanding the sequences satisfying the recurrence. If the roots r_1, r_2, ... are all distinct, then each solution to the recurrence takes the form

$$a_n = k_1 r_1^n + k_2 r_2^n + \cdots + k_d r_d^n,$$

where the coefficients k_i are determined in order to fit the initial conditions of the recurrence. When the same roots occur multiple times, the terms in this formula corresponding to the second and later occurrences of the same root are multiplied by increasing powers of n. For instance, if the characteristic polynomial can be factored as $(x-r)^3$, with the same root r occurring three times, then the solution would take the form

$$a_n = k_1 r^n + k_2 n r^n + k_3 n^2 r^n.$$

As well as the Fibonacci numbers, other constant-recursive sequences include the Lucas numbers and Lucas sequences, the Jacobsthal numbers, the Pell numbers and more generally the solutions to Pell's equation.

For order 1, the recurrence

$$a_n = r a_{n-1}$$

has the solution $a_n = r^n$ with $a_0 = 1$ and the most general solution is $a_n = kr^n$ with $a_0 = k$. The characteristic polynomial equated to zero (the characteristic equation) is simply $t - r = 0$.

Solutions to such recurrence relations of higher order are found by systematic means, often using the fact that $a_n = r^n$ is a solution for the recurrence exactly when $t = r$ is a root of the characteristic polynomial. This can be approached directly or using generating functions (formal power series) or matrices.

Consider, for example, a recurrence relation of the form

$$a_n = A a_{n-1} + B a_{n-2}.$$

When does it have a solution of the same general form as $a_n = r^n$? Substituting this guess (ansatz) in the recurrence relation, we find that

$$r^n = A r^{n-1} + B r^{n-2}$$

must be true for all $n > 1$.

Dividing through by r^{n-2}, we get that all these equations reduce to the same thing:

$$r^2 = Ar + B,$$

$$r^2 - Ar - B = 0,$$

which is the characteristic equation of the recurrence relation. Solve for r to obtain the two roots λ_1, λ_2: these roots are known as the characteristic roots or eigenvalues of the characteristic equation. Different solutions are obtained depending on the nature of the roots: If these roots are distinct, we have the general solution

$$a_n = C\lambda_1^n + D\lambda_2^n$$

while if they are identical (when $A^2 + 4B = 0$), we have

$$a_n = C\lambda^n + Dn\lambda^n$$

This is the most general solution; the two constants C and D can be chosen based on two given initial conditions a_0 and a_1 to produce a specific solution.

In the case of complex eigenvalues (which also gives rise to complex values for the solution parameters C and D), the use of complex numbers can be eliminated by rewriting the solution in trigonometric form. In this case we can write the eigenvalues as $\lambda_1, \lambda_2 = \alpha \pm \beta i$. Then it can be shown that

$$a_n = C\lambda_1^n + D\lambda_2^n.$$

can be rewritten as

$$a_n = 2M^n \left(E\cos(\theta n) + F\sin(\theta n) \right) = 2GM^n \cos(\theta n - \delta),$$

where

$$M = \sqrt{\alpha^2 + \beta^2} \quad \cos(\theta) = \tfrac{\alpha}{M} \quad \sin(\theta) = \tfrac{\beta}{M}$$
$$C, D = E \mp Fi$$
$$G = \sqrt{E^2 + F^2} \quad \cos(\delta) = \tfrac{E}{G} \quad \sin(\delta) = \tfrac{F}{G}$$

Here E and F (or equivalently, G and δ) are real constants which depend on the initial conditions. Using

$$\lambda_1 + \lambda_2 = 2\alpha = A,$$

$$\lambda_1 \cdot \lambda_2 = \alpha^2 + \beta^2 = -B,$$

one may simplify the solution given above as

$$a_n = (-B)^{\frac{n}{2}} \left(E\cos(\theta n) + F\sin(\theta n) \right),$$

where a_1 and a_2 are the initial conditions and

$$E = \frac{-Aa_1 + a_2}{B}$$

$$F = -i\frac{A^2a_1 - Aa_2 + 2a_1B}{B\sqrt{A^2 + 4B}}$$

$$\theta = \arccos\left(\frac{A}{2\sqrt{-B}}\right)$$

In this way there is no need to solve for λ_1 and λ_2.

In all cases—real distinct eigenvalues, real duplicated eigenvalues, and complex conjugate eigenvalues—the equation is stable (that is, the variable a converges to a fixed value [specifically, zero]) if and only if *both* eigenvalues are smaller than one in absolute value. In this second-order case, this condition on the eigenvalues can be shown to be equivalent to $|A| < 1 - B < 2$, which is equivalent to $|B| < 1$ and $|A| < 1 - B$.

The equation in the above example was homogeneous, in that there was no constant term. If one starts with the non-homogeneous recurrence

$$b_n = Ab_{n-1} + Bb_{n-2} + K$$

with constant term K, this can be converted into homogeneous form as follows: The steady state is found by setting $b_n = b_{n-1} = b_{n-2} = b^*$ to obtain

$$b^* = \frac{K}{1 - A - B}.$$

Then the non-homogeneous recurrence can be rewritten in homogeneous form as

$$[b_n - b^*] = A[b_{n-1} - b^*] + B[b_{n-2} - b^*],$$

which can be solved as above.

The stability condition stated above in terms of eigenvalues for the second-order case remains valid for the general n^{th}-order case: the equation is stable if and only if all eigenvalues of the characteristic equation are less than one in absolute value.

Given a homogeneous linear recurrence relation with constant coefficients of order d, let $p(t)$ be the characteristic polynomial (also "auxiliary polynomial")

$$t^d - c_1t^{d-1} - c_2t^{d-2} - \cdots - c_d = 0$$

such that each c_i corresponds to each c_i in the original recurrence relation. Suppose λ is a root of $p(t)$ having multiplicity r. This is to say that $(t-\lambda)^r$ divides $p(t)$. The following two properties hold:

1. Each of the r sequences $\lambda^n, n\lambda^n, n^2\lambda^n, \ldots, n^{r-1}\lambda^n$ satisfies the recurrence relation.

2. Any sequence satisfying the recurrence relation can be written uniquely as a linear combination of solutions constructed in part 1 as λ varies over all distinct roots of $p(t)$.

As a result of this theorem a homogeneous linear recurrence relation with constant coefficients can be solved in the following manner:

1. Find the characteristic polynomial $p(t)$.

2. Find the roots of $p(t)$ counting multiplicity.

3. Write a_n as a linear combination of all the roots (counting multiplicity as shown in the theorem above) with unknown coefficients b_i.

$$a_n = \left(b_1\lambda_1^n + b_2 n\lambda_1^n + b_3 n^2\lambda_1^n + \cdots + b_r n^{r-1}\lambda_1^n\right) + \cdots + \left(b_{d-q+1}\lambda_*^n + \cdots + b_d n^{q-1}\lambda_*^n\right)$$

This is the general solution to the original recurrence relation. (q is the multiplicity of λ_*.)

4. Equate each $a_0, a_1, \ldots, a_d$ from part 3 (plugging in $n = 0, \ldots, d$ into the general solution of the recurrence relation) with the known values $a_0, a_1, \ldots, a_d$ from the original recurrence relation. However, the values a_n from the original recurrence relation used do not usually have to be contiguous: excluding exceptional cases, just d of them are needed (i.e., for an original homogeneous linear recurrence relation of order 3 one could use the values a_0, a_1, a_4). This process will produce a linear system of d equations with d unknowns. Solving these equations for the unknown coefficients $b_1, b_2, \ldots, b_d$ of the general solution and plugging these values back into the general solution will produce the particular solution to the original recurrence relation that fits the original recurrence relation's initial conditions (as well as all subsequent values $a_0, a_1, a_2, \ldots$ of the original recurrence relation).

The method for solving linear differential equations is similar to the method above—the "intelligent guess" (ansatz) for linear differential equations with constant coefficients is $e^{\lambda x}$ where λ is a complex number that is determined by substituting the guess into the differential equation.

This is not a coincidence. Considering the Taylor series of the solution to a linear differential equation:

$$\sum_{n=0}^{\infty} \frac{f^{(n)}(a)}{n!}(x-a)^n$$

it can be seen that the coefficients of the series are given by the nth derivative of $f(x)$ evaluated at the point a. The differential equation provides a linear difference equation relating these coefficients.

This equivalence can be used to quickly solve for the recurrence relationship for the coefficients in the power series solution of a linear differential equation.

The rule of thumb (for equations in which the polynomial multiplying the first term is non-zero at zero) is that:

$$y^{[k]} \rightarrow f[n+k]$$

and more generally

$$x^m * y^{[k]} \rightarrow n(n-1)...(n-m+1)f[n+k-m]$$

Example: The recurrence relationship for the Taylor series coefficients of the equation:

$$(x^2 + 3x - 4)y^{[3]} - (3x + 1)y^{[2]} + 2y = 0$$

is given by

$$n(n-1)f[n+1] + 3nf[n+2] - 4f[n+3] - 3nf[n+1] - f[n+2] + 2f[n] = 0$$

or

$$-4f[n+3] + 2nf[n+2] + n(n-4)f[n+1] + 2f[n] = 0.$$

This example shows how problems generally solved using the power series solution method taught in normal differential equation classes can be solved in a much easier way.

Example: The differential equation

$$ay'' + by' + cy = 0$$

has solution

$$y = e^{ax}.$$

The conversion of the differential equation to a difference equation of the Taylor coefficients is

$$af[n+2] + bf[n+1] + cf[n] = 0.$$

It is easy to see that the nth derivative of e^{ax} evaluated at 0 is a^n

Solving via Linear Algebra

A linearly recursive sequence y of order n

$$y_{n+k} - c_{n-1}y_{n-1+k} - c_{n-2}y_{n-2+k} + \cdots - c_0 y_k = 0$$

is identical to

$$y_n = c_{n-1}y_{n-1} + c_{n-2}y_{n-2} + \cdots + c_0 y_0.$$

Expanded with n-1 identities of kind $y_{n-k} = y_{n-k}$, this n-th order equation is translated into a matrix difference equation system of n first-order linear equations,

$$\vec{y}_n = \begin{bmatrix} y_n \\ y_{n-1} \\ \vdots \\ \vdots \\ y_1 \end{bmatrix} = \begin{bmatrix} c_{n-1} & c_{n-2} & \cdots & \cdots & c_0 \\ 1 & 0 & \cdots & \cdots & 0 \\ 0 & \ddots & \ddots & & \vdots \\ \vdots & & \ddots & \ddots & \vdots \\ 0 & \cdots & 0 & 1 & 0 \end{bmatrix} \begin{bmatrix} y_{n-1} \\ y_{n-2} \\ \vdots \\ \vdots \\ y_0 \end{bmatrix} = C\,\vec{y}_{n-1} = C^n \vec{y}_0.$$

Observe that the vector $\vec{y}_n$ can be computed by n applications of the companion matrix, C, to the initial state vector, y_0. Thereby, n-th entry of the sought sequence y, is the top component of $\vec{y}_n$, $y_n = \vec{y}_n[n]$.

Eigendecomposition, $\vec{y}_n = C^n \vec{y}_0 = c_1 \lambda_1^n \vec{e}_1 + c_2 \lambda_2^n \vec{e}_2 + \cdots + c_n \lambda_n^n \vec{e}_n$ into eigenvalues, $\lambda_1, \lambda_2, \ldots, \lambda_n$, and eigenvectors, $\vec{e}_1, \vec{e}_2, \ldots, \vec{e}_n$, is used to compute $\vec{y}_n$. Thanks to the crucial fact that system C time-shifts every eigenvector, e, by simply scaling its components λ times,

$$C\vec{e}_i = \lambda_i \vec{e}_i = C \begin{bmatrix} e_{i,n} \\ e_{i,n-1} \\ \vdots \\ e_{i,1} \end{bmatrix} = \begin{bmatrix} \lambda_i e_{i,n} \\ \lambda_i e_{i,n-1} \\ \vdots \\ \lambda_i e_{i,1} \end{bmatrix}$$

that is, time-shifted version of eigenvector, e, has components λ times larger, the eigenvector components are powers of λ, $\vec{e}_i = \begin{bmatrix} \lambda_i^{n-1} & \cdots & \lambda_i^2 & \lambda_i & 1 \end{bmatrix}^T$, and, thus, recurrent homogeneous linear equation solution is a combination of exponential functions, $\vec{y}_n = \sum_1^n c_i \lambda_i^n \vec{e}_i$. The components c_i can be determined out of initial conditions:

$$\vec{y}_0 = \begin{bmatrix} y_0 \\ y_{-1} \\ \vdots \\ y_{-n+1} \end{bmatrix} = \sum_{i=1}^n c_i \lambda_i^0 \vec{e}_i = \begin{bmatrix} \vec{e}_1 & \vec{e}_2 & \cdots & \vec{e}_n \end{bmatrix} \begin{bmatrix} c_1 \\ c_2 \\ \cdots \\ c_n \end{bmatrix} = E \begin{bmatrix} c_1 \\ c_2 \\ \cdots \\ c_n \end{bmatrix}$$

Solving for coefficients,

$$\begin{bmatrix} c_1 \\ c_2 \\ \cdots \\ c_n \end{bmatrix} = E^{-1} \vec{y}_0 = \begin{bmatrix} \lambda_1^{n-1} & \lambda_2^{n-1} & \cdots & \lambda_n^{n-1} \\ \vdots & \vdots & \ddots & \vdots \\ \lambda_1 & \lambda_2 & \cdots & \lambda_n \\ 1 & 1 & \cdots & 1 \end{bmatrix}^{-1} \begin{bmatrix} y_0 \\ y_{-1} \\ \vdots \\ y_{-n+1} \end{bmatrix}.$$

This also works with arbitrary boundary conditions $\underbrace{y_a, y_b, \ldots}_{n}$, not necessary the initial ones,

$$
\begin{bmatrix} y_a \\ y_b \\ \vdots \end{bmatrix} = \begin{bmatrix} \vec{y}_a[n] \\ \vec{y}_b[n] \\ \vdots \end{bmatrix} = \begin{bmatrix} \sum_{i=1}^{n} c_i \lambda_i^a \vec{e}_i[n] \\ \sum_{i=1}^{n} c_i \lambda_i^b \vec{e}_i[n] \\ \vdots \end{bmatrix} = \begin{bmatrix} \sum_{i=1}^{n} c_i \lambda_i^a \lambda_i^{n-1} \\ \sum_{i=1}^{n} c_i \lambda_i^b \lambda_i^{n-1} \\ \vdots \end{bmatrix} =
$$

$$
= \begin{bmatrix} \sum c_i \lambda_i^{a+n-1} \\ \sum c_i \lambda_i^{b+n-1} \\ \vdots \end{bmatrix} = \begin{bmatrix} \lambda_1^{a+n-1} & \lambda_2^{a+n-1} & \cdots & \lambda_n^{a+n-1} \\ \lambda_1^{b+n-1} & \lambda_2^{b+n-1} & \cdots & \lambda_n^{b+n-1} \\ \vdots & \vdots & \vdots & \ddots & \vdots \end{bmatrix} \begin{bmatrix} c_1 \\ c_2 \\ \vdots \\ c_n \end{bmatrix}.
$$

This description is really no different from general method above, however it is more succinct. It also works nicely for situations like

$$
\begin{cases} a_n = a_{n-1} - b_{n-1} \\ b_n = 2a_{n-1} + b_{n-1}. \end{cases}
$$

where there are several linked recurrences.

Solving with z-transforms

Certain difference equations - in particular, linear constant coefficient difference equations - can be solved using z-transforms. The z-transforms are a class of integral transforms that lead to more convenient algebraic manipulations and more straightforward solutions. There are cases in which obtaining a direct solution would be all but impossible, yet solving the problem via a thoughtfully chosen integral transform is straightforward.

Solving Non-homogeneous Linear Recurrence Relations with Constant Coefficients

If the recurrence is non-homogeneous, a particular solution can be found by the method of un-determined coefficients and the solution is the sum of the solution of the homogeneous and the particular solutions. Another method to solve an non-homogeneous recurrence is the method of *symbolic differentiation*. For example, consider the following recurrence:

$$
a_{n+1} = a_n + 1
$$

This is an non-homogeneous recurrence. If we substitute $n \mapsto n+1$, we obtain the recurrence

$$
a_{n+2} = a_{n+1} + 1
$$

Subtracting the original recurrence from this equation yields

$$a_{n+2} - a_{n+1} = a_{n+1} - a_n$$

or equivalently

$$a_{n+2} = 2a_{n+1} - a_n$$

This is a homogeneous recurrence, which can be solved by the methods explained above. In general, if a linear recurrence has the form

$$a_{n+k} = \lambda_{k-1}a_{n+k-1} + \lambda_{k-2}a_{n+k-2} + \cdots + \lambda_1 a_{n+1} + \lambda_0 a_n + p(n)$$

where $\lambda_0, \lambda_1, \ldots, \lambda_{k-1}$ are constant coefficients and $p(n)$ is the inhomogeneity, then if $p(n)$ is a polynomial with degree r, then this non-homogeneous recurrence can be reduced to a homogeneous recurrence by applying the method of symbolic differencing r times.

If

$$P(x) = \sum_{n=0}^{\infty} p_n x^n$$

is the generating function of the inhomogeneity, the generating function

$$A(x) = \sum_{n=0}^{\infty} a(n) x^n$$

of the non-homogeneous recurrence

$$a_n = \sum_{i=1}^{s} c_i a_{n-i} + p_n, \quad n \geq n_r,$$

with constant coefficients c_i is derived from

$$\left(1 - \sum_{i=1}^{s} c_i x^i\right) A(x) = P(x) + \sum_{n=0}^{n_r-1} [a_n - p_n] x^n - \sum_{i=1}^{s} c_i x^i \sum_{n=0}^{n_r-i-1} a_n x^n.$$

If $P(x)$ is a rational generating function, $A(x)$ is also one. The case discussed above, where $p_n = K$ is a constant, emerges as one example of this formula, with $P(x) = K/(1-x)$. Another example, the recurrence $a_n = 10a_{n-1} + n$ with linear inhomogeneity, arises in the definition of the schizophrenic numbers. The solution of homogeneous recurrences is incorporated as $p = P = 0$.

Solving First-order Non-homogeneous Recurrence Relations with Variable Coefficients

Moreover, for the general first-order non-homogeneous linear recurrence relation with variable coefficients:

$$a_{n+1} = f_n a_n + g_n, \qquad f_n \neq 0,$$

there is also a nice method to solve it:

$$a_{n+1} - f_n a_n = g_n$$

$$\frac{a_{n+1}}{\prod\limits_{k=0}^{n} f_k} - \frac{f_n a_n}{\prod\limits_{k=0}^{n} f_k} = \frac{g_n}{\prod\limits_{k=0}^{n} f_k}$$

$$\frac{a_{n+1}}{\prod\limits_{k=0}^{n} f_k} - \frac{a_n}{\prod\limits_{k=0}^{n-1} f_k} = \frac{g_n}{\prod\limits_{k=0}^{n} f_k}$$

Let

$$A_n = \frac{a_n}{\prod\limits_{k=0}^{n-1} f_k},$$

Then

$$A_{n+1} - A_n = \frac{g_n}{\prod\limits_{k=0}^{n} f_k}$$

$$\sum_{m=0}^{n-1} (A_{m+1} - A_m) = A_n - A_0 = \sum_{m=0}^{n-1} \frac{g_m}{\prod\limits_{k=0}^{m} f_k}$$

$$\frac{a_n}{\prod\limits_{k=0}^{n-1} f_k} = A_0 + \sum_{m=0}^{n-1} \frac{g_m}{\prod\limits_{k=0}^{m} f_k}$$

$$a_n = \left(\prod_{k=0}^{n-1} f_k \right) \left(A_0 + \sum_{m=0}^{n-1} \frac{g_m}{\prod\limits_{k=0}^{m} f_k} \right)$$

If we apply the formula to $a_{n+1} = (1 + h f_{nh}) a_n + h g_{nh}$ and take the limit h→0, we get the formula for first order linear differential equations with variable coefficients; the sum becomes an integral, and the product becomes the exponential function of an integral.

Solving general Homogeneous Linear Recurrence Relations

Many homogeneous linear recurrence relations may be solved by means of the generalized hypergeometric series. Special cases of these lead to recurrence relations for the orthogonal polynomials, and many special functions. For example, the solution to

$$J_{n+1} = -J_n - J_{n-1}$$

is given by

$$J_n = J_n(z),$$

the Bessel function, while

$$(b-n)M_{n-1} + (2n-b-z)M_n - nM_{n+1} = 0$$

is solved by

$$M_n = M(n,b;z)$$

the confluent hypergeometric series.

Solving First-order Rational Difference Equations

A first order rational difference equation has the form $w_{t+1} = \frac{aw_t+b}{cw_t+d}$. Such an equation can be solved by writing w_t as a nonlinear transformation of another variable x_t which itself evolves linearly. Then standard methods can be used to solve the linear difference equation in x_t.

Stability

Stability of Linear Higher-order Recurrences

The linear recurrence of order d,

$$a_n = c_1 a_{n-1} + c_2 a_{n-2} + \cdots + c_d a_{n-d},$$

has the characteristic equation

$$\lambda^d - c_1 \lambda^{d-1} - c_2 \lambda^{d-2} - \quad -c \, \lambda^0 = 0.$$

The recurrence is stable, meaning that the iterates converge asymptotically to a fixed value, if and only if the eigenvalues (i.e., the roots of the characteristic equation), whether real or complex, are all less than unity in absolute value.

Stability of Linear First-order Matrix Recurrences

In the first-order matrix difference equation

$$[x_t - x^*] = A[x_{t-1} - x^*]$$

with state vector x and transition matrix A, x converges asymptotically to the steady state vector x^* if and only if all eigenvalues of the transition matrix A (whether real or complex) have an absolute value which is less than 1.

Stability of Nonlinear First-order Recurrences

Consider the nonlinear first-order recurrence

$$x_n = f(x_{n-1}).$$

This recurrence is locally stable, meaning that it converges to a fixed point x^* from points sufficiently close to x^*, if the slope of f in the neighborhood of x^* is smaller than unity in absolute value: that is,

$$|f'(x^*)| < 1.$$

A nonlinear recurrence could have multiple fixed points, in which case some fixed points may be locally stable and others locally unstable; for continuous f two adjacent fixed points cannot both be locally stable.

A nonlinear recurrence relation could also have a cycle of period k for $k > 1$. Such a cycle is stable, meaning that it attracts a set of initial conditions of positive measure, if the composite function

$$g(x) := f \circ f \circ \cdots \circ f(x)$$

with f appearing k times is locally stable according to the same criterion:

$$|g'(x^*)| < 1,$$

where x^* is any point on the cycle.

In a chaotic recurrence relation, the variable x stays in a bounded region but never converges to a fixed point or an attracting cycle; any fixed points or cycles of the equation are unstable.

Relationship to Differential Equations

When solving an ordinary differential equation numerically, one typically encounters a recurrence relation. For example, when solving the initial value problem

$$y'(t) = f(t, y(t)), \quad y(t_0) = y_0,$$

with Euler's method and a step size h, one calculates the values

$$y_0 = y(t_0), \quad y_1 = y(t_0 + h), \quad y_2 = y(t_0 + 2h), \ldots$$

by the recurrence

$$y_{n+1} = y_n + hf(t_n, y_n).$$

Systems of linear first order differential equations can be discretized exactly analytically.

Applications

Biology

Some of the best-known difference equations have their origins in the attempt to model population dynamics. For example, the Fibonacci numbers were once used as a model for the growth of a rabbit population.

The logistic map is used either directly to model population growth, or as a starting point for more detailed models of population dynamics. In this context, coupled difference equations are often used to model the interaction of two or more populations. For example, the Nicholson-Bailey model for a host-parasite interaction is given by

$$N_{t+1} = \lambda N_t e^{-aP_t}$$
$$P_{t+1} = N_t (1 - e^{-aP_t}),$$

with N_t representing the hosts, and P_t the parasites, at time t.

Integrodifference equations are a form of recurrence relation important to spatial ecology. These and other difference equations are particularly suited to modeling univoltine populations.

Computer Science

Recurrence relations are also of fundamental importance in analysis of algorithms. If an algorithm is designed so that it will break a problem into smaller subproblems (divide and conquer), its running time is described by a recurrence relation.

A simple example is the time an algorithm takes to find an element in an ordered vector with n elements, in the worst case.

A naive algorithm will search from left to right, one element at a time. The worst possible scenario is when the required element is the last, so the number of comparisons is n.

A better algorithm is called binary search. However, it requires a sorted vector. It will first check if the element is at the middle of the vector. If not, then it will check if the middle element is greater or lesser than the sought element. At this point, half of the vector can be discarded, and the algorithm can be run again on the other half. The number of comparisons will be given by

$$c_1 = 1$$
$$c_n = 1 + c_{n/2}$$

which will be close to $\log_2(n)$.

Digital Signal Processing

In digital signal processing, recurrence relations can model feedback in a system, where outputs at one time become inputs for future time. They thus arise in infinite impulse response (IIR) digital filters.

For example, the equation for a "feedforward" IIR comb filter of delay T is:

$$y_t = (1-\alpha)x_t + \alpha y_{t-T}$$

Where x_t is the input at time t, y_t is the output at time t, and α controls how much of the delayed signal is fed back into the output. From this we can see that

$$y_t = (1-\alpha)x_t + \alpha((1-\alpha)x_{t-T} + \alpha y_{t-2T})$$
$$y_t = (1-\alpha)x_t + (\alpha - \alpha^2)x_{t-T} + \alpha^2 y_{t-2T})$$

etc.

Economics

Recurrence relations, especially linear recurrence relations, are used extensively in both theoretical and empirical economics. In particular, in macroeconomics one might develop a model of various broad sectors of the economy (the financial sector, the goods sector, the labor market, etc.) in which some agents' actions depend on lagged variables. The model would then be solved for current values of key variables (interest rate, real GDP, etc.) in terms of exogenous variables and lagged endogenous variables.

This section contains the applications of formal power series to solving recurrence relations. Let us try to understand it using the following examples.

Example: 1. Determine a formula for the numbers a(n)'s, where a(n)'s satisfy the recurrence relation a(n) = 3 a(n − 1) + 2n, for n ≥ 1 with a(0) = 1.

Solution: Define $A(x) = \sum_{n \geq 0} a(n)x^n$. Then using Example, one has

$$A(x) = \sum_{n \geq 0} a(n)x^n = a_0 + \sum_{n \geq 1} a(n)x^n = 1 + \sum_{n \geq 1}(3a(n-1)+2n)x^n.$$

$$= 3x\sum_{n \geq 1} a(n-1)x^{n-1} + 2\sum_{n \geq 1} nx^n + 1 = 3xA(x) + 2\frac{x}{(1-x)^2} + 1.$$

$$So, \ A(x) = \frac{1+x^2}{(1-3x)(1-x)^2} = \frac{5}{2(1-3x)} - \frac{1}{2(1-x)} - \frac{1}{(1-x)^2}. \ Thus,$$

$$a(n) = \left[x^n\right]A(x) = \frac{5}{2}3^n - \frac{1}{2} - (n+1) = \frac{5.3^n - 1}{2} - (n+1).$$

2. Determine a generating function for the numbers f (n) that satisfy the recurrence relation

f (n) = f (n − 1) + f (n − 2),   for n ≥ 2   with f (0) = 1 and f (1) = 1.

Hence or otherwise find a formula for the numbers f (n).

Solution: Define $F(x) = \sum_{n \geq 0} f(n)x^n$. Then one has

$$F(x) = \sum_{n \geq 0} f(n)x^n = 1 + x\sum_{n \geq 2}(f(n-1) + f(n-2))x^n$$

$$= 1 + x + x\sum_{n \geq 2} f(n-1)x^{n-1} + x^2\sum_{n \geq 2} f(n-2)x^{n-2} = 1 + xF(x) + x^2 F(x).$$

Therefore, $F(x) = \dfrac{1}{1-x-x^2}$. Let $\alpha = \dfrac{1+\sqrt{5}}{2}$ and $\beta = \dfrac{1-\sqrt{5}}{2}$. Then it can be checked that $(1-\alpha x)(1-\beta x) = 1 - x - x^2$ and

$$F(x) = \frac{1}{\sqrt{5}}\left(\frac{\alpha}{1-\alpha x} - \frac{\beta}{1-\beta x}\right) = \frac{1}{\sqrt{5}}\left(\sum_{n \geq 0}\alpha^{n+1}x^n - \sum_{n \geq 0}\beta^{n+1}x^n\right).$$

Therefore,

$$f(n) = [x^n]F(x) = \frac{1}{\sqrt{5}}\sum_{n \geq 0}(\alpha^{n+1} - \beta^{n+1}).$$

As $\beta < 0$ and $|\beta| < 1$, we observe that $f(n) \approx \dfrac{1}{\sqrt{5}}\left(\dfrac{1+\sqrt{5}}{2}\right)^{n+1}$.

Remark: The numbers f (n), for n ≥ 0 are called Fibonacci numbers. It is related with the following problem: Suppose a couple bought a pair of rabbits (each one year old) in the year 2001. If a pair of rabbits starts giving birth to a pair of rabbits as soon as they grow 2 years old, determine the number of rabbits the couple will have in the year 2025.

3. Determine a formula for the numbers a(n)'s, where a(n)'s satisfy the recurrence relation a(n) = 3 a(n − 1) + 4a(n − 2), for n ≥ 2 with a(0) = 1 and a(1) = c, a constant.

Solution: Define $A(x) = \sum_{n \geq 0} a(n)x^n$. Then

$$A(x) = \sum_{n \geq 0} a(n)x^n = a_0 + a_1 x + \sum_{n \geq 2} a(n)x^n = 1 + cx + \sum_{n \geq 2}(3a(n-1) + 4a(n-2))x^n$$

$$= 1 + cx + 3x\sum_{n \geq 2} a(n-1)x^{n-1} + 4x^2\sum_{n \geq 2} a_{n-2}x^{n-2}$$

$$= 1 + cx + 3x(A(x) - a_0) + 4x^2 A(x).$$

So, $A(x) = \dfrac{1 + (c-3)x}{(1 - 3x - 4x^2)} = \dfrac{1 + (c-3)x}{(1-x)(1-4x)}.$

(a) If $c = 4$ then $A(x) = \dfrac{1}{1-4x}$ and hence $a_n = [x^n]A(x) = 4^n..$

(b) If $c \neq 4$ then $A(x) = \dfrac{1+c}{5}\cdot\dfrac{1}{1-4x} + \dfrac{4-c}{5}\cdot\dfrac{1}{1+x}$ and hence $a_n = [x^n]A(x) = \dfrac{(1+c)4^n}{5} + \dfrac{(-1)^n(4-c)}{5}.$

4. Determine a sequence, $\{a(n) \in \mathbb{R} : n \geq 0\}$, such that $a_0 = 1$ and $\sum_{k=0}^{n} a(k)a(n-k) = \binom{n+2}{2}$, for all n ≥ 1.

Solution: Define $A(x) = \sum_{n\geq 0} a(n)x^n$. Then, using the Cauchy product, one has

$$A(x)^2 = \sum_{n\geq 0}\left(\sum_{k=0}^{n} a(k)a(n-k)\right)x^n = \sum_{n\geq 0}\binom{n+2}{2}x^n = \frac{1}{(1-x)^3}.$$

Hence, $A(x) = \frac{1}{(1-x)^{3/2}}$ and thus $a(n) = (-1)^n\binom{-3/2}{n} = \frac{3\cdot 5\cdot 7\cdots(2n+1)}{2^n n!}$, for all n ≥ 1.

5. Determine a generating function for the numbers $f(n,m), n, m \in \mathbb{Z}, n, m \geq 0$ that satisfy

$$f(n,m) = f(n-1,m) + f(n-1,m-1), (n,m) \neq (0,\ 0)$$
$$f(n,0) = 1, \text{ for all } n\geq 0 \text{ and } f(0,m) = 0, \text{ for all } m > 0.$$

Hence or otherwise, find a formula for the numbers f (n, m).

Solution: Note that in the above recurrence relation, the value of m need not be ≤ n.

Method 1: Define $F_n(x) = \sum_{m\geq 0} f(n,m)x^m$. Then, for n ≥ 1, earlier equation gives

$$F_n(x) = \sum_{m\geq 0} f(n,m)x^m = \sum_{m\geq 0}\left(f(n-1,m) + f(n-1,m-1)\right)x^m$$
$$= \sum_{m\geq 0} f(n-1,m)x^m + \sum_{m\geq 0} f(n-1,m-1)x^m$$
$$= F_{n-1}(x) + xF_{n-1}(x) = (1+x)F_{n-1}(x) = \cdots = (1+x)^n F_0(x).$$

Now, using the initial conditions, $F_0(x) = 1$ and hence $F_n(x) = (1+x)^n$. Thus,

$$f(n,m) = [x^m](1+x)^n = \binom{n}{m} \text{ if } 0\leq m\leq n \text{ and } \quad f(n,m) = 0, \text{ for } \quad m > n.$$

Method 2: Define $G_m(y) = \sum_{n\geq 0} f(n,\ m)y^n$. Then, for m ≥ 1, earlier equation gives

$$G_m(y) = \sum_{n\geq 0} f(n,m)y^n = \sum_{n\geq 0}(f(n-1,m) + f(n-1,m-1))y^n$$
$$= \sum_{n\geq 0} f(n-1,\ m)y^n + \sum_{n\geq 0} f(n-1,\ m-1)y^n$$
$$= yG_m(y) + yG_{m-1}(y).$$

Therefore, $G_m(y) = \frac{1}{1-y}G_{m-1}(y)$. Now, using initial conditions, $G_0(y) = \frac{1}{1-y}$ and hence

$G_m(y) = \frac{y^m}{(1-y)^{m+1}}$. Thus, $f(n.m) = [y^n]\frac{y^m}{(1-y)^{m+1}} = [y^{n-m}]\frac{1}{(1-y)^{m+1}} = \binom{n}{m}$, whenever 0 ≤ m ≤ n and

f (n, m) = 0, for m > n.

Example: Determine a generating function for the numbers $S(n,m), n, m \in \mathbb{Z}, n, m \geq 0$ that satisfy

$$f(n,m) = f(n-1,m) + f(n-1,m-1), (n,m) \neq (0,\ 0)$$
$$f(n,0) = 1, \text{ for all } n \geq 0 \text{ and } f(0,m) = 0, \text{ for all } m > 0.$$

Hence or otherwise find a formula for the numbers S(n, m).

Solution: Define $G_m(y) = \sum_{n \geq 0} S(n,m) y^n$. Then, for m ≥ 1, earlier equation gives

$$G_m(y) = \sum_{n \geq 0} S(n,m) y^n = \sum_{n \geq 0} (mS(n-1,m) + S(n-1,m-1)) y^n$$
$$= m \sum_{n \geq 0} S(n-1,m) y^n + \sum_{n \geq 0} S(n-1,m-1) y^n$$
$$= myG_m(y) + yG_{m-1}(y).$$

Therefore, $G_m(y) = \dfrac{y}{1-my} G_{m-1}(y)$. Using initial conditions, $G_0(y) = 1$ and hence

$$G_m(y) = \frac{y^m}{(1-y)(1-2y)\cdots(1-my)} = y^m \sum_{k=1}^{m} \frac{\alpha_k}{1-ky},$$

where $\alpha_k = \dfrac{(-1)^{m-k} k^m}{k!(m-k)!}$, for 1 ≤ k ≤ m. Thus,

$$S(n,m) = \left[y^n \right]\left(y^m \sum_{k=1}^{m} \frac{\alpha_k}{1-ky} \right) = \sum_{k=1}^{m} \left[y^{n-m} \right] \frac{\alpha_k}{1-ky}$$
$$= \sum_{k=1}^{m} \alpha_k k^{n-m} = \sum_{k=1}^{m} \frac{(-1)^{m-k} k^n}{k!(m-k)!}$$
$$= \frac{1}{m!} \sum_{k=1}^{m} (-1)^{m-k} k^n \binom{m}{k} = \frac{1}{m!} \sum_{k=1}^{m} (-1)^k (m-k)^n \binom{m}{k}.$$

Therefore, $S(n,m) = \dfrac{1}{m!} \sum_{k=1}^{m} (-1)^k (m-k)^n \binom{m}{k}$ and $m! S(n,m) = \sum_{k=1}^{m} (-1)^k (m-k)^n \binom{m}{k}$.

This identity is generally known as the Stirling's Identity.

Observation:

(a) $H_n(x) = \sum_{m \geq 0} S(n,m) x^m$ is not considered. But verify that

$$H_n(x) = (x + xD)^n \cdot 1 \text{ as } \quad H_0(x) = 1.$$

Therefore, $H_1(x) = x, H_2(x) = x + x^2, \cdots$. Hence, it is difficult to obtain a general formula for its coefficients. But it is helpful in showing that the numbers S(n, m),

for fixed n, first increase and then decrease (commonly called unimodal). The same holds for the sequence of binomial coefficients $\left\{ \binom{n}{m}, m = 0, 1, ..., n \right\}$.

(b) Since there is no restriction on the non-negative integers n and m, the expression Equation (3) is also valid for n < m. But, in this case, we know that S(n, m) = 0. Hence, verify that
$$\sum_{k=1}^{m} \frac{(-1)^{m-k} k^{n-1}}{(k-1)!(m-k)!} = 0 \text{ , whenever n < m.}$$

2. Bell Numbers: For a positive integer n, the nth Bell number, denoted b(n), is the number of partitions of the set {1, 2, . . . , n}. Therefore, by definition, $b(n) = \sum_{m=1}^{m} S(n,m)$,for n ≥ 1 and by convention, b(0) = 1. Thus, for n ≥ 1,

$$b(n) = \sum_{m=1}^{m} S(n,m) = \sum_{m\geq 1} S(n,m) = \sum_{m\geq 1} \sum_{k=1}^{m} \frac{(-1)^{m-k} k^{n-1}}{(k-1)!(m-k)!}$$
$$= \sum_{k\geq 1} \frac{k^n}{k!} \sum_{m\geq k} \frac{(1-)^{m-k}}{(m-k)!} = \frac{1}{e} \sum_{k\geq 1} \frac{k^n}{k!}.$$

Note that the above equation is valid even for n = 0. Also, observe that b(n) has terms of the form $\frac{k^n}{k!}$ and hence we compute its exponential generating function.

Thus, if $B(x) = \sum_{n\geq 0} b(n) \frac{x^n}{n!}$ then

$$B(x) = 1 + \sum_{n\geq 1} b(n) \frac{x^n}{n!} = 1 + \sum_{n\geq 1} \left(\frac{1}{e} \sum_{k\geq 1} \frac{k^n}{k!} \right) \frac{x^n}{n!}$$

$$= 1 + \frac{1}{e} \sum_{k\geq 1} \frac{1}{k!} \sum_{n\geq 1} k^n \frac{x^n}{n!} = 1 + \frac{1}{e} \sum_{k\geq 1} \frac{1}{k!} \sum_{n\geq 1} \frac{(kx)^n}{n!}$$

$$= 1 + \frac{1}{e} \sum_{k\geq 1} \frac{1}{k!} (e^{kx} - 1) = 1 + \frac{1}{e} \sum_{k\geq 1} \left(\frac{(e^x)^k}{k!} - \frac{1}{k!} \right)$$

$$= 1 + \frac{1}{e} (e^{e^x} - 1 - (e-1)) = e^{e^x} - 1.$$

Recall that $e^{e^x} - 1$ is a valid formal power series. Now, let us derive the recurrence relation for b(n)'s. Taking the natural logarithm on both the sides of above equation, one has $Ln\left(\sum_{n\geq 0} b(n) \frac{x^n}{n!} \right) = e^x - 1$.

Now, differentiation with respect to x gives $\frac{1}{\sum_{n\geq 0} b(n) \frac{x^n}{n!}} \cdot \sum_{n\geq 0} b(n) \frac{x^{n-1}}{(n-1)!} = e^x$. Therefore, after cross multiplication and a multiplica-tion with x, implies

$$\sum_{n\geq 1} \frac{b(n) x^n}{(n-1)!} = xe^x \sum_{n\geq 0} b(n) \frac{x^n}{n!} = x \left(\sum_{m\geq 0} \frac{x^m}{m!} \right) \cdot \left(\sum_{n\geq 0} b(n) \frac{x^n}{n!} \right).$$

Thus,

$$\frac{b(n)}{(n-1)!} = \left[x^n\right]\frac{b(n)x^n}{(n-1)!} = \left[x^n\right]x\left(\sum_{m\geq0}\frac{x^m}{m!}\right)\cdot\left(\sum_{n\geq0}b(n)\frac{x^n}{n!}\right) = \sum_{m=0}^{n-1}\frac{1}{(n-1-m)!}\cdot\frac{b(m)}{m!}$$

Hence, it follows that $b(n) = \sum_{m=0}^{n-1}\binom{n-1}{m}b(m)$, for n ≥ 1, with b(0) = 1.

Example: Determine the number of ways of arranging n pairs of parentheses (left and right) such that at any stage the number of right parentheses is always less than or equal to the number of left parentheses.

Solution: Recall that this number equals C_n, the nth Catalan number. Let us obtain a recurrence relation for these numbers and use it to get a formula for C_n's.

Let P_n denote the arrangements of those n pairs of parentheses that satisfy "at any stage, the number of left parentheses is always greater than or equal to the number of right parentheses". Then $|P_n| = C_n$, for all n ≥ 1. Also, let Q_n denote those elements of P_n for which, "at the 2k-th stage, for k < n, the number of left parentheses is strictly greater than the number of right parentheses".

We now claim that $|Q_1| = 1$ and $|Q_n| = |P_{n-1}|$, for n ≥ 2.

Clearly $|Q_1| = 1$. Note that, for n ≥ 2, any element of Q_n, necessarily starts with two left parentheses and ends with two right parentheses. So, if we remove the first left parenthesis and the last right parenthesis from each element of Q_n then one obtains an element of P_{n-1}. In a similar way, if we add one left parenthesis at the beginning and a right parenthesis at the end of an element of P_{n-1}, one obtains an element of Q_n. Hence, there is one-to-one correspondence between the set Q_n and P_{n-1}. Thus, $|Q_n| = |P_{n-1}| = C_{n-1}$.

Let n ≥ 2 and consider an element of P_n. Then, for some k, 1 ≤ k ≤ n, the first k pairs of parentheses will have the property that the number of left parentheses is strictly greater than the number of right parentheses, for 1 ≤ ℓ < k, i.e., they form an element of Q_k and the remaining (n − k) pairs of parentheses will form an element of P_{n-k}. Hence, if $|P_0| = |Q_1| = 1$, one has

$$C_n = |P_n| = \sum_{k=1}^{n}|Q_k||P_{n-k}| = \sum_{k=1}^{n}|P_{k-1}||P_{n-k}| = \sum_{k=1}^{n}C_{k-1}C_{n-k},\ for\ n\geq2.$$ Now, define $C(x) = \sum_{n\geq0}C_nx^n$. Then

$$C(x) = \sum_{n\geq0}C_nx^n = 1+\sum_{n\geq1}C_nx^n = 1+\sum_{n\geq1}\left(\sum_{k=1}^{n}C_{k-1}C_{n-k}\right)x^n$$

$$= 1+x\left(\sum_{k\geq1}C_{k-1}x^{k-1}\sum_{n\geq k}C_{n-k}x^{n-k}\right) = 1+x\left(C(x)\sum_{k\geq1}C_{k-1}x^{k-1}\right)$$

$$= 1+x(C(x))^2.$$

Thus, $xC(x)^2 - C(x) + 1 = 0$ and $C(x) = \dfrac{1\pm\sqrt{1-4x}}{2x}$. Therefore, using $C_0 = 1$, one obtains

$C(x) = \dfrac{1-\sqrt{1-4x}}{2x}$. Hence,

$$C_n = \left[x^n\right] C(x) = \frac{1}{2} \cdot \left[x^{n+1}\right] (1 - \sqrt{1-4x})$$

$$= -\frac{1}{2} \cdot \frac{\frac{1}{2} \left(\frac{1}{2}-1\right) \left(\frac{1}{2}-2\right) \cdots \left(\frac{1}{2}-n\right)}{(n+1)!} (-4)^{n+1}$$

$$= 2(-4)^n \cdot \frac{1 \cdot (-1) \cdot (-3) \cdot (-5) \cdots (1-2n)}{2^{n+1}(n+1)!} = 2^n \frac{1 \cdot 3 \cdot 5 \cdots (2n-1)}{(n+1)!}$$

$$= \frac{1}{n+1} \binom{2n}{n}, \text{ the } n^{th} \text{ Catalan Number.}$$

The ideas learnt in the previous sections will be used to get closed form expressions for sums arising out of binomial coefficients.

Example: Find a closed form expression for the numbers $a(n) = \sum_{k \geq 0} \binom{k}{n-k}$.

Solution: Define $A(x) = \sum_{n \geq 0} a(n)x^n$. Then

$$A(x) = \sum_{n \geq 0} a(n)x^n = \sum_{n \geq 0} \left(\sum_{k \geq 0} \binom{k}{n-k}\right) x^n = \sum_{k \geq 0} \left(\sum_{n \geq 0} \binom{k}{n-k} x^n\right)$$

$$= \sum_{k \geq 0} x^k \left(\sum_{n \geq k} \binom{k}{n-k} x^{n-k}\right) = \sum_{k \geq 0} x^k (1+x)^k = \sum_{k \geq 0} x^k (x(1+x))^k = \frac{1}{1-x(1-x)}.$$

Therefore, Example implies $a(n) = \left[x^n\right] A(x) = \left[x^n\right] \frac{1}{1-x(1+x)} = F_n$, the n-th Fibonacci number.

2. Find a closed form expression for the polynomials $a(n,x) = \sum_{k=0}^{\left\lfloor \frac{n}{2} \right\rfloor} \binom{n-k}{k} (-1)^k x^{n-2k}$.

Solution: Define $A(x,y) = \sum_{n \geq 0} a(n,x)y^n$. Then

$$A(x,y) = \sum_{n \geq 0} a(n,x)y^n = \sum_{n \geq 0} \left(\sum_{k=0}^{\left\lfloor \frac{n}{2} \right\rfloor} \binom{n-k}{k} (-1)^k x^{n-2k}\right) y^n$$

$$= \sum_{k \geq 0} (-1)^k y^{2k} \left(\sum_{n \geq 2k} \binom{n-k}{k} (xy)^{n-2k}\right)$$

$$= \sum_{k \geq 0} (-1)^k y^{2k} (xy)^{-k} \left(\sum_{t \geq k} \binom{t}{k} (xy)^t\right)$$

$$= \sum_{k \geq 0} (-y^2)^k (xy)^{-k} \frac{(xy)^k}{(1-xy)^{k+1}} = \frac{1}{1-xy} \cdot \sum_{k \geq 0} \left(\frac{(-y^2)}{1-xy}\right)^k$$

$$= \frac{1}{1-xy} \cdot \frac{1}{1 - \frac{-y^2}{1-xy}} = \frac{1}{1-xy+y^2} = \frac{1}{(1-\alpha y)(1-\beta y)},$$

where $\alpha = \dfrac{x+\sqrt{x^2-4}}{2}$ and $\beta = \dfrac{x-\sqrt{x^2-4}}{2}$. Thus,

$$a(n,x)=\left[y^n\right]A(x,y)=\left[y^n\right]\frac{1}{1-xy+y^2}=\left[y^n\right]\frac{1}{\alpha-\beta}\left(\frac{\alpha}{1-\alpha y}-\frac{\beta}{1-\beta y}\right)$$

$$=\frac{1}{\alpha-\beta}\left(\alpha^{n+1}-\beta^{n+1}\right)$$

$$=\frac{1}{\sqrt{x^2-4}}\left(\left(\frac{x+\sqrt{x^2-4}}{2}\right)^{n+1}-\left(\frac{x-\sqrt{x^2-4}}{2}\right)^{n+1}\right).$$

Since α and β are the roots of $y^2-xy+1=0$, $\alpha^2=\alpha x-1$ and $\beta^2=\beta x-1$. Therefore, verify that the $a(n,x)$'s satisfy the recurrence relation $a(n,x)=xa(n-1,x)-a(n-2,x)$, for $n\geq 2$, with initial conditions $a(0,x)=1$ and $a(1,x)=x$.

Let A = (a_{ij}) be an n × n matrix, with a_{ij} = 1, whenever $|i-j|$ = 1 and 0, otherwise. Then A is an adjacency matrix of a tree T on n vertices, say 1, 2, . . . , n with the vertex i being adjacent to i + 1, for $1 \leq i \leq n-1$. It can be verified that if $a(n,x)=det(xI_n-A)$, the characteristic polynomial of A, then $a(n,x)$'s satisfy the above recurrence relation. The polynomials $a(n,2x)$'s are also known as Chebyshev's polynomial of second kind.

Let us now substitute different values for x to obtain different expressions and then use them to get binomial identities.

1. Let $x = z + \dfrac{1}{z}$. Then $\sqrt{x^2-4} = z - \dfrac{1}{z}$ and we obtain $a\left(n, z+\dfrac{1}{z}\right)=\dfrac{z^{2n+2}-1}{(z^2-1)z^n}$. Hence, we have

$$\sum_{k=0}^{\left\lfloor\frac{n}{2}\right\rfloor}\binom{n-k}{k}(-1)^k\left(z+\frac{1}{z}\right)^{n-2k}=\frac{z^{2n+2}-1}{(z^2-1)z^n}.$$ Or equivalently,

$$\sum_{k=0}^{\left\lfloor\frac{n}{2}\right\rfloor}\binom{n-k}{k}(-1)^k(z^2+1)^{n-2k}z^{2k}=\frac{z^{2n+2}-1}{z^2-1}.$$

2. Writing x in place of z², we obtain the following identity.

$$\sum_{k=0}^{\left\lfloor\frac{n}{2}\right\rfloor}\binom{n-k}{k}(-1)^k(x+1)^{n-2k}x^k=\frac{x^{n+1}+1}{x-1}=\sum_{k=0}^{n}x^k.$$

3. Hence, equating the coefficient of x^m in the above equation gives the identity

$$\sum_{k=0}^{\left\lfloor\frac{n}{2}\right\rfloor}(-1)^k\binom{n-k}{k}\binom{n-2k}{m-k}=\begin{cases}1, & \text{if } 0\leq m\leq n;\\ 0, & \text{otherwise.}\end{cases}$$

4. Substituting $x = 1$ gives $\displaystyle\sum_{k=0}^{\left\lfloor \frac{n}{2} \right\rfloor} (-1)^k \binom{n-k}{k} 2^{n-2k} = n+1$.

Example: Determine the generating function for the numbers $\displaystyle a(n,y) = \sum_{k \geq 0} \binom{n+k}{2k} y^k$.

Solution: Define $\displaystyle A(x,y) = \sum_{n \geq 0} a(n,y) x^n$. Then

$$A(x,y) = \sum_{n \geq 0} \left(\sum_{k \geq 0} \binom{n+k}{2k} y^k \right) x^n = \sum_{k \geq 0} \left(\frac{y}{x} \right)^k \sum_{n \geq k} \binom{n+k}{2k} x^{n+k}$$

$$= \sum_{k \geq 0} \left(\frac{y}{x} \right)^k \frac{x^{2k}}{(1-x)^{2k+1}} = \frac{1}{1-x} \sum_{k \geq 0} \left(\frac{yx}{(1-x)^2} \right)^k$$

$$= \frac{1-x}{(1-x)^2 - xy}.$$

1. Verify that if we substituting y = −2 then

$$\sum_{k \geq 0} \binom{n+k}{2k} (-2)^k = \left[x^n \right] A(x,-2) = \left[x^n \right] \frac{1-x}{(1+x^2)} = (-1)^{\lceil n/2 \rceil}.$$

2. Verify that if we substituting y = −4 then

$$\sum_{k \geq 0} \binom{n+k}{2k} (-4)^k = \left[x^n \right] A(x,-4) = \left[x^n \right] \frac{1-x}{(1+x)^2} = (-1)^n (2n+1).$$

3. Let $\displaystyle f(n) = \sum_{k \geq 0} \binom{n+k}{2k} 2^{n-k}$ and let $\displaystyle F(z) = \sum_{n \geq 0} f(n) z^n$. Then verify that

$$F(z) = A\left(2z, \frac{1}{2} \right) = \frac{1-2z}{(1-z)(1-4z)} = \frac{2}{3} \cdot \frac{1}{1-4z} + \frac{1}{3} \cdot \frac{1}{1-z}.$$

Hence, $\displaystyle f(n) = \left[z^n \right] F(z) = \frac{2 \cdot 4^n}{3} + \frac{1}{3} = \frac{2^{2n+1} + 1}{3}$.

References

- Grimaldi, Ralph P. (1994), Discrete and Combinatorial Mathematics: An Applied Introduction (3rd ed.), ISBN 978-0-201-54983-6

- Björklund, A.; Husfeldt, T.; Koivisto, M. (2009), "Set partitioning via inclusion–exclusion", SIAM Journal on Computing, 39 (2): 546–563, doi:10.1137/070683933

- Berstel, Jean; Reutenauer, Christophe (2011). Noncommutative rational series with applications. Encyclopedia of Mathematics and Its Applications. 137. Cambridge: Cambridge University Press. ISBN 978-0-521-19022-0. Zbl 1250.68007

- Thomas H. Cormen, Charles E. Leiserson, Ronald L. Rivest, and Clifford Stein. Introduction to Algorithms, Second Edition. MIT Press and McGraw-Hill, 1990. ISBN 0-262-03293-7

- Wang, Xiang-Sheng; Wong, Roderick (2012). "Asymptotics of orthogonal polynomials via recurrence relations". Anal. Appl. 10 (2): 215–235. doi:10.1142/S0219530512500108

- Jacques, Ian (2006). Mathematics for Economics and Business (Fifth ed.). Prentice Hall. pp. 551–568. ISBN 0-273-70195-9

An Integrated Study of Mathematical Groups

When two elements combine to form a third element, a group is formed. Groups are mostly used along with the use of polynomial equations. The aspects elucidated in this chapter are of vital importance, and provide a better understanding of discrete mathematics.

Group (Mathematics)

The manipulations of this Rubik's Cube form the Rubik's Cube group.

In mathematics, a group is an algebraic structure consisting of a set of elements equipped with an operation that combines any two elements to form a third element. The operation satisfies four conditions called the group axioms, namely closure, associativity, identity and invertibility. One of the most familiar examples of a group is the set of integers together with the addition operation, but the abstract formalization of the group axioms, detached as it is from the concrete nature of any particular group and its operation, applies much more widely. It allows entities with highly diverse mathematical origins in abstract algebra and beyond to be handled in a flexible way while retaining their essential structural aspects. The ubiquity of groups in numerous areas within and outside mathematics makes them a central organizing principle of contemporary mathematics.

Groups share a fundamental kinship with the notion of symmetry. For example, a symmetry group encodes symmetry features of a geometrical object: the group consists of the set of transformations that leave the object unchanged and the operation of combining two such transformations by performing one after the other. Lie groups are the symmetry groups used in the Standard Model of particle physics; Poincaré groups, which are also Lie groups, can express the physical symmetry underlying special relativity; and point groups are used to help understand symmetry phenomena in molecular chemistry.

The concept of a group arose from the study of polynomial equations, starting with Évariste Galois in the 1830s. After contributions from other fields such as number theory and geometry, the group notion was generalized and firmly established around 1870. Modern group theory—an active mathematical discipline—studies groups in their own right. To explore groups, mathematicians have devised various notions to break groups into smaller, better-understandable pieces, such as subgroups, quotient groups and simple groups. In addition to their abstract properties, group theorists also study the different ways in which a group can be expressed concretely, both from a point of view of representation theory (that is, through the representations of the group) and of computational group theory. A theory has been developed for finite groups, which culminated with the classification of finite simple groups, completed in 2004. Since the mid-1980s, geometric group theory, which studies finitely generated groups as geometric objects, has become a particularly active area in group theory.

Definition and Illustration

First Example: The Integers

One of the most familiar groups is the set of integers Z which consists of the numbers

..., −4, −3, −2, −1, 0, 1, 2, 3, 4, ..., together with addition.

The following properties of integer addition serve as a model for the abstract group axioms given in the definition below:

- For any two integers a and b, the sum $a + b$ is also an integer. That is, addition of integers always yields an integer. This property is known as *closure* under addition.

- For all integers a, b and c, $(a + b) + c = a + (b + c)$. Expressed in words, adding a to b first, and then adding the result to c gives the same final result as adding a to the sum of b and c, a property known as *associativity*.

- If a is any integer, then $0 + a = a + 0 = a$. Zero is called the *identity element* of addition because adding it to any integer returns the same integer.

- For every integer a, there is an integer b such that $a + b = b + a = 0$. The integer b is called the *inverse element* of the integer a and is denoted $-a$.

The integers, together with the operation +, form a mathematical object belonging to a broad class sharing similar structural aspects. To appropriately understand these structures as a collective, the following abstract definition is developed.

Definition

The axioms for a group are short and natural. Yet somehow hidden behind these axioms is the monster simple group, a huge and extraordinary mathematical object, which appears to rely on numerous bizarre coincidences to exist. The axioms for groups give no obvious hint that anything like this exists.

A group is a set, G, together with an operation • (called the *group law* of G) that combines any two elements a and b to form another element, denoted $a \bullet b$ or ab. To qualify as a group, the set and operation, $(G, \bullet)$, must satisfy four requirements known as the *group axioms*:

Closure

> For all a, b in G, the result of the operation, $a \bullet b$, is also in G.

Associativity

> For all a, b and c in G, $(a \bullet b) \bullet c = a \bullet (b \bullet c)$.

Identity element

> There exists an element e in G such that, for every element a in G, the equation $e \bullet a = a \bullet e = a$ holds. Such an element is unique, and thus one speaks of *the* identity element.

Inverse element

> For each a in G, there exists an element b in G, commonly denoted a^{-1} (or $-a$, if the operation is denoted "+"), such that $a \bullet b = b \bullet a = e$, where e is the identity element.

The result of an operation may depend on the order of the operands. In other words, the result of combining element a with element b need not yield the same result as combining element b with element a; the equation

$$a \bullet b = b \bullet a$$

may not always be true. This equation always holds in the group of integers under addition, because $a + b = b + a$ for any two integers (commutativity of addition). Groups for which the commutativity equation $a \bullet b = b \bullet a$ always holds are called *abelian groups* (in honor of Niels Henrik Abel). The symmetry group described in the following section is an example of a group that is not abelian.

The identity element of a group G is often written as 1 or 1_G, a notation inherited from the multiplicative identity. If a group is abelian, then one may choose to denote the group operation by + and the identity element by 0; in that case, the group is called an additive group. The identity element can also be written as *id*.

The set G is called the *underlying set* of the group $(G, \bullet)$. Often the group's underlying set G is used as a short name for the group $(G, \bullet)$. Along the same lines, shorthand expressions such as "a subset of the group G" or "an element of group G" are used when what is actually meant is "a subset of the underlying set G of the group $(G, \bullet)$" or "an element of the underlying set G of the group $(G, \bullet)$". Usually, it is clear from the context whether a symbol like G refers to a group or to an underlying set.

Second Example: A Symmetry Group

Two figures in the plane are congruent if one can be changed into the other using a combination of rotations, reflections, and translations. Any figure is congruent to itself. However, some figures are congruent to themselves in more than one way, and these extra congruences are called symmetries. A square has eight symmetries. These are:

The elements of the symmetry group of the square (D₄). Vertices are identified by color or number.			
id (keeping it as it is)	r_1 (rotation by 90° clockwise)	r_2 (rotation by 180° clockwise)	r_3 (rotation by 270° clockwise)
f_v (vertical reflection)	f_h (horizontal reflection)	f_d (diagonal reflection)	f_c (counter-diagonal reflection)

- the identity operation leaving everything unchanged, denoted id;

- rotations of the square around its center by 90° clockwise, 180° clockwise, and 270° clockwise, denoted by r_1, r_2 and r_3, respectively;

- reflections about the vertical and horizontal middle line (f_h and f_v), or through the two diagonals (f_d and f_c).

These symmetries are represented by functions. Each of these functions sends a point in the square to the corresponding point under the symmetry. For example, r_1 sends a point to its rotation 90° clockwise around the square's center, and f_h sends a point to its reflection across the square's vertical middle line. Composing two of these symmetry functions gives another symmetry function. These symmetries determine a group called the dihedral group of degree 4 and denoted D_4. The underlying set of the group is the above set of symmetry functions, and the group operation is function composition. Two symmetries are combined by composing them as functions, that is, applying the first one to the square, and the second one to the result of the first application. The result of performing first a and then b is written symbolically *from right to left* as

$b \cdot a$ ("apply the symmetry b after performing the symmetry a").

The right-to-left notation is the same notation that is used for composition of functions.

The group table on the right lists the results of all such compositions possible. For example, rotating by 270° clockwise (r_3) and then reflecting horizontally (f_h) is the same as performing a reflection along the diagonal (f_d). Using the above symbols, highlighted in blue in the group table:

$$f_h \cdot r_3 = f_d.$$

•	id	r_1	r_2	r_3	f_v	f_h	f_d	f_c
id	id	r_1	r_2	r_3	f_v	f_h	f_d	f_c
r_1	r_1	r_2	r_3	id	f_c	f_d	f_v	f_h
r_2	r_2	r_3	id	r_1	f_h	f_v	f_c	f_d
r_3	r_3	id	r_1	r_2	f_d	f_c	f_h	f_v
f_v	f_v	f_d	f_h	f_c	id	r_2	r_1	r_3
f_h	f_h	f_c	f_v	f_d	r_2	id	r_3	r_1
f_d	f_d	f_h	f_c	f_v	r_3	r_1	id	r_2
f_c	f_c	f_v	f_d	f_h	r_1	r_3	r_2	id

Group table of D_4

The elements id, r_1, r_2, and r_3 form a subgroup, highlighted in red (upper left region). A left and right coset of this subgroup is highlighted in green (in the last row) and yellow (last column), respectively.

Given this set of symmetries and the described operation, the group axioms can be understood as follows:

1. The closure axiom demands that the composition $b \bullet a$ of any two symmetries a and b is also a symmetry. Another example for the group operation is

$$r_3 \bullet f_h = f_c,$$

i.e., rotating 270° clockwise after reflecting horizontally equals reflecting along the counter-diagonal (f_c). Indeed every other combination of two symmetries still gives a symmetry, as can be checked using the group table.

2. The associativity constraint deals with composing more than two symmetries: Starting with three elements a, b and c of D_4, there are two possible ways of using these three symmetries in this order to determine a symmetry of the square. One of these ways is to first compose a and b into a single symmetry, then to compose that symmetry with c. The other way is to first compose b and c, then to compose the resulting symmetry with a. The associativity condition

$$(a \bullet b) \bullet c = a \bullet (b \bullet c)$$

means that these two ways are the same, i.e., a product of many group elements can be simplified in any grouping. For example, $(f_d \bullet f_v) \bullet r_2 = f_d \bullet (f_v \bullet r_2)$ can be checked using the group table.

$(f_d \bullet f_v) \bullet r_2 = r_3 \bullet r_2 = r_1$, which equals

$f_d \bullet (f_v \bullet r_2) = f_d \bullet f_h = r_1$.

While associativity is true for the symmetries of the square and addition of numbers, it is not true for all operations. For instance, subtraction of numbers is not associative: $(7 - 3) - 2 = 2$ is not the same as $7 - (3 - 2) = 6$.

3. The identity element is the symmetry id leaving everything unchanged: for any symmetry a, performing id after a (or a after id) equals a, in symbolic form,

    id $\bullet$ a = a,

    a $\bullet$ id = a.

4. An inverse element undoes the transformation of some other element. Every symmetry can be undone: each of the following transformations—identity id, the reflections f_h, f_v, f_d, f_c and the 180° rotation r_2—is its own inverse, because performing it twice brings the square back to its original orientation. The rotations r_3 and r_1 are each other's inverses, because rotating 90° and then rotation 270° (or vice versa) yields a rotation over 360° which leaves the square unchanged. In symbols,

    f_h $\bullet$ f_h = id,

    r_3 $\bullet$ r_1 = r_1 $\bullet$ r_3 = id.

In contrast to the group of integers above, where the order of the operation is irrelevant, it does matter in D_4: f_h $\bullet$ r_1 = f_c but r_1 $\bullet$ f_h = f_d. In other words, D_4 is not abelian, which makes the group structure more difficult than the integers introduced first.

History

The modern concept of an abstract group developed out of several fields of mathematics. The original motivation for group theory was the quest for solutions of polynomial equations of degree higher than 4. The 19th-century French mathematician Évariste Galois, extending prior work of Paolo Ruffini and Joseph-Louis Lagrange, gave a criterion for the solvability of a particular polynomial equation in terms of the symmetry group of its roots (solutions). The elements of such a Galois group correspond to certain permutations of the roots. At first, Galois' ideas were rejected by his contemporaries, and published only posthumously. More general permutation groups were investigated in particular by Augustin Louis Cauchy. Arthur Cayley's *On the theory of groups, as depending on the symbolic equation $\theta^n = 1$* (1854) gives the first abstract definition of a finite group.

Geometry was a second field in which groups were used systematically, especially symmetry groups as part of Felix Klein's 1872 Erlangen program. After novel geometries such as hyperbolic and projective geometry had emerged, Klein used group theory to organize them in a more coherent way. Further advancing these ideas, Sophus Lie founded the study of Lie groups in 1884.

The third field contributing to group theory was number theory. Certain abelian group structures had been used implicitly in Carl Friedrich Gauss' number-theoretical work *Disquisitiones Arithmeticae* (1798), and more explicitly by Leopold Kronecker. In 1847, Ernst Kummer made early attempts to prove Fermat's Last Theorem by developing groups describing factorization into prime numbers.

The convergence of these various sources into a uniform theory of groups started with Camille Jordan's *Traité des substitutions et des équations algébriques* (1870). Walther von Dyck (1882) introduced the idea of specifying a group by means of generators and relations, and was also the first to give an axiomatic definition of an "abstract group", in the terminology of the time. As of the

20th century, groups gained wide recognition by the pioneering work of Ferdinand Georg Frobenius and William Burnside, who worked on representation theory of finite groups, Richard Brauer's modular representation theory and Issai Schur's papers. The theory of Lie groups, and more generally locally compact groups was studied by Hermann Weyl, Élie Cartan and many others. Its algebraic counterpart, the theory of algebraic groups, was first shaped by Claude Chevalley (from the late 1930s) and later by the work of Armand Borel and Jacques Tits.

The University of Chicago's 1960–61 Group Theory Year brought together group theorists such as Daniel Gorenstein, John G. Thompson and Walter Feit, laying the foundation of a collaboration that, with input from numerous other mathematicians, led to the classification of finite simple groups, with the final step taken by Aschbacher and Smith in 2004. This project exceeded previous mathematical endeavours by its sheer size, in both length of proof and number of researchers. Research is ongoing to simplify the proof of this classification. These days, group theory is still a highly active mathematical branch, impacting many other fields.

Elementary Consequences of the Group Axioms

Basic facts about all groups that can be obtained directly from the group axioms are commonly subsumed under *elementary group theory*. For example, repeated applications of the associativity axiom show that the unambiguity of

$$a \bullet b \bullet c = (a \bullet b) \bullet c = a \bullet (b \bullet c)$$

generalizes to more than three factors. Because this implies that parentheses can be inserted anywhere within such a series of terms, parentheses are usually omitted.

The axioms may be weakened to assert only the existence of a left identity and left inverses. Both can be shown to be actually two-sided, so the resulting definition is equivalent to the one given above.

Uniqueness of Identity Element and Inverses

Two important consequences of the group axioms are the uniqueness of the identity element and the uniqueness of inverse elements. There can be only one identity element in a group, and each element in a group has exactly one inverse element. Thus, it is customary to speak of *the* identity, and *the* inverse of an element.

To prove the uniqueness of an inverse element of a, suppose that a has two inverses, denoted b and c, in a group $(G, \bullet)$. Then

$b = b \bullet e$	as e is the identity element
$= b \bullet (a \bullet c)$	because c is an inverse of a, so $e = a \bullet c$
$= (b \bullet a) \bullet c$	by associativity, which allows to rearrange the parentheses
$= e \bullet c$	since b is an inverse of a, i.e., $b \bullet a = e$
$= c$	for e is the identity element

The term b on the first line above and the c on the last are equal, since they are connected by a chain of equalities. In other words, there is only one inverse element of a. Similarly, to prove that the identity element of a group is unique, assume G is a group with two identity elements e and f. Then $e = e \cdot f = f$, hence e and f are equal.

Division

In groups, the existence of inverse elements implies that division is possible: given elements a and b of the group G, there is exactly one solution x in G to the equation $x \cdot a = b$, namely $b \cdot a^{-1}$. In fact, we have

$$(b \cdot a^{-1}) \cdot a = b \cdot (a^{-1} \cdot a) = b \cdot e = b.$$

Uniqueness results by multiplying the two sides of the equation $x \cdot a = b$ by a^{-1}. The element $b \cdot a^{-1}$, often denoted b / a, is called the *right quotient* of b by a, or the result of the *right division* of b by a.

Similarly there is exactly one solution y in G to the equation $a \cdot y = b$, namely $y = a^{-1} \cdot b$. This solution is the *left quotient* of b by a, and is sometimes denoted $a \setminus b$.

In general b / a and $a \setminus b$ may be different, but, if the group operation is commutative (that is, if the group is abelian), they are equal. In this case, the group operation is often denoted as an addition, and one talks of *subtraction* and *difference* instead of division and quotient.

A consequence of this is that multiplication by a group element g is a bijection. Specifically, if g is an element of the group G, the function (mathematics) from G to itself that maps $h \in G$ to $g \cdot h$ is a bijection. This function is called the *left translation* by g. Similarly, the *right translation* by g is the bijection from G to itself, that maps h to $h \cdot g$. If G is abelian, the left and the right translation by a group element are the same.

Basic Concepts

The following sections use mathematical symbols such as $X = \{x, y, z\}$ to denote a set X containing elements x, y, and z, or alternatively $x \in X$ to restate that x is an element of X. The notation $f : X \to Y$ means f is a function assigning to every element of X an element of Y.

To understand groups beyond the level of mere symbolic manipulations as above, more structural concepts have to be employed. There is a conceptual principle underlying all of the following notions: to take advantage of the structure offered by groups (which sets, being "structureless", do not have), constructions related to groups have to be *compatible* with the group operation. This compatibility manifests itself in the following notions in various ways. For example, groups can be related to each other via functions called group homomorphisms. By the mentioned principle, they are required to respect the group structures in a precise sense. The structure of groups can also be understood by breaking them into pieces called subgroups and quotient groups. The principle of "preserving structures"—a recurring topic in mathematics throughout—is an instance of working in a category, in this case the category of groups.

Group Homomorphisms

Group homomorphisms are functions that preserve group structure. A function $a : G \to H$ between two groups $(G, \cdot)$ and $(H, *)$ is called a *homomorphism* if the equation

$$a(g \cdot k) = a(g) * a(k)$$

holds for all elements g, k in G. In other words, the result is the same when performing the group operation after or before applying the map a. This requirement ensures that $a(1_G) = 1_H$, and also $a(g)^{-1} = a(g^{-1})$ for all g in G. Thus a group homomorphism respects all the structure of G provided by the group axioms.

Two groups G and H are called *isomorphic* if there exist group homomorphisms $a: G \rightarrow H$ and $b: H \rightarrow G$, such that applying the two functions one after another in each of the two possible orders gives the identity functions of G and H. That is, $a(b(h)) = h$ and $b(a(g)) = g$ for any g in G and h in H. From an abstract point of view, isomorphic groups carry the same information. For example, proving that $g \cdot g = 1_G$ for some element g of G is equivalent to proving that $a(g) * a(g) = 1_H$, because applying a to the first equality yields the second, and applying b to the second gives back the first.

Subgroups

Informally, a *subgroup* is a group H contained within a bigger one, G. Concretely, the identity element of G is contained in H, and whenever h_1 and h_2 are in H, then so are $h_1 \cdot h_2$ and h_1^{-1}, so the elements of H, equipped with the group operation on G restricted to H, indeed form a group.

In the example above, the identity and the rotations constitute a subgroup $R = \{\text{id}, r_1, r_2, r_3\}$, highlighted in red in the group table above: any two rotations composed are still a rotation, and a rotation can be undone by (i.e., is inverse to) the complementary rotations 270° for 90°, 180° for 180°, and 90° for 270° (note that rotation in the opposite direction is not defined). The subgroup test is a necessary and sufficient condition for a nonempty subset H of a group G to be a subgroup: it is sufficient to check that $g^{-1}h \in H$ for all elements g, $h \in H$. Knowing the subgroups is important in understanding the group as a whole.

Given any subset S of a group G, the subgroup generated by S consists of products of elements of S and their inverses. It is the smallest subgroup of G containing S. In the introductory example above, the subgroup generated by r_2 and f_v consists of these two elements, the identity element id and $f_h = f_v \cdot r_2$. Again, this is a subgroup, because combining any two of these four elements or their inverses (which are, in this particular case, these same elements) yields an element of this subgroup.

Cosets

In many situations it is desirable to consider two group elements the same if they differ by an element of a given subgroup. For example, in D_4 above, once a reflection is performed, the square never gets back to the r_2 configuration by just applying the rotation operations (and no further reflections), i.e., the rotation operations are irrelevant to the question whether a reflection has been performed. Cosets are used to formalize this insight: a subgroup H defines left and right cosets, which can be thought of as translations of H by arbitrary group elements g. In symbolic terms, the *left* and *right* cosets of H containing g are

$gH = \{g \cdot h : h \in H\}$ and $Hg = \{h \cdot g : h \in H\}$, respectively.

The left cosets of any subgroup H form a partition of G; that is, the union of all left cosets is equal to G and two left cosets are either equal or have an empty intersection. The first case $g_1H = g_2H$

happens precisely when $g_1^{-1} \cdot g_2 \in H$, i.e., if the two elements differ by an element of H. Similar considerations apply to the right cosets of H. The left and right cosets of H may or may not be equal. If they are, i.e., for all g in G, $gH = Hg$, then H is said to be a *normal subgroup*.

In D_4, the introductory symmetry group, the left cosets gR of the subgroup R consisting of the rotations are either equal to R, if g is an element of R itself, or otherwise equal to $U = f_c R = \{f_c, f_v,$ $f_d, f_h\}$ (highlighted in green). The subgroup R is also normal, because $f_c R = U = Rf_c$ and similarly for any element other than f_c. (In fact, in the case of D_4, observe that all such cosets are equal, such that $f_h R = f_v R = f_d R = f_c R$.)

Quotient Groups

In some situations the set of cosets of a subgroup can be endowed with a group law, giving a *quotient group* or *factor group*. For this to be possible, the subgroup has to be normal. Given any normal subgroup N, the quotient group is defined by

$$G \, / \, N = \{gN, g \in G\}, \text{"}G \text{ modulo } N\text{".}$$

This set inherits a group operation (sometimes called coset multiplication, or coset addition) from the original group G: $(gN) \cdot (hN) = (gh)N$ for all g and h in G. This definition is motivated by the idea (itself an instance of general structural considerations outlined above) that the map $G \to G \, / \, N$ that associates to any element g its coset gN be a group homomorphism, or by general abstract considerations called universal properties. The coset $eN = N$ serves as the identity in this group, and the inverse of gN in the quotient group is $(gN)^{-1} = (g^{-1})N$.

Group table of the quotient group $D_4 \, / \, R$		
•	**R**	**U**
R	*R*	*U*
U	*U*	*R*

The elements of the quotient group $D_4 \, / \, R$ are R itself, which represents the identity, and $U = f_v R$. The group operation on the quotient is shown. For example, $U \cdot U = f_v R \cdot f_v R = (f_v \cdot f_v)R = R$. Both the subgroup $R = \{id, r_1, r_2, r_3\}$, as well as the corresponding quotient are abelian, whereas D_4 is not abelian. Building bigger groups by smaller ones, such as D_4 from its subgroup R and the quotient $D_4 \, / \, R$ is abstracted by a notion called semidirect product.

Quotient groups and subgroups together form a way of describing every group by its *presentation*: any group is the quotient of the free group over the *generators* of the group, quotiented by the subgroup of *relations*. The dihedral group D_4, for example, can be generated by two elements r and f (for example, $r = r_1$, the right rotation and $f = f_v$ the vertical (or any other) reflection), which means that every symmetry of the square is a finite composition of these two symmetries or their inverses. Together with the relations

$$r^4 = f^2 = (r \cdot f)^2 = 1,$$

the group is completely described. A presentation of a group can also be used to construct the Cayley graph, a device used to graphically capture discrete groups.

Sub- and quotient groups are related in the following way: a subset H of G can be seen as an injective map $H \to G$, i.e., any element of the target has at most one element that maps to it. The counterpart to injective maps are surjective maps (every element of the target is mapped onto), such as the canonical map $G \to G/N$. Interpreting subgroup and quotients in light of these homomorphisms emphasizes the structural concept inherent to these definitions alluded to in the introduction. In general, homomorphisms are neither injective nor surjective. Kernel and image of group homomorphisms and the first isomorphism theorem address this phenomenon.

Examples and Applications

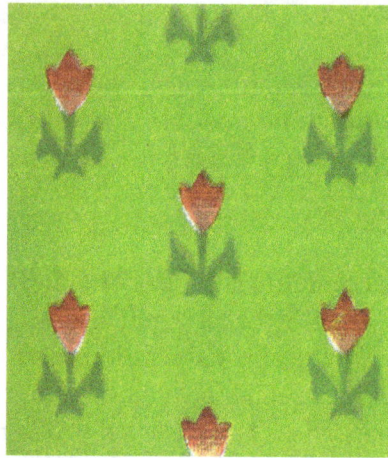

A periodic wallpaper pattern gives rise to a wallpaper group.

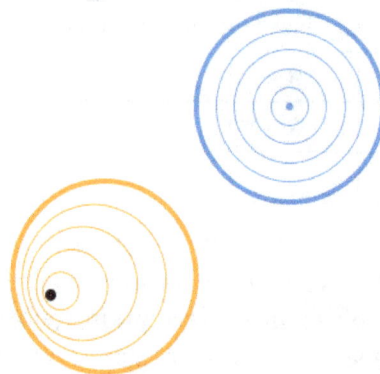

The fundamental group of a plane minus a point (bold) consists of loops around the missing point.
This group is isomorphic to the integers.

Examples and applications of groups abound. A starting point is the group Z of integers with addition as group operation, introduced above. If instead of addition multiplication is considered, one obtains multiplicative groups. These groups are predecessors of important constructions in abstract algebra.

Groups are also applied in many other mathematical areas. Mathematical objects are often examined by associating groups to them and studying the properties of the corresponding groups. For example, Henri Poincaré founded what is now called algebraic topology by introducing the fundamental group. By means of this connection, topological properties such as proximity and continuity translate into properties of groups. For example, elements of the fundamental group are represented by loops. The second image at the right shows some loops in a plane minus a point. The blue loop is considered null-homotopic (and thus irrelevant), because it can be continuously shrunk to a point. The presence of the hole prevents the orange loop from being shrunk to a point. The fundamental group of the plane with a point deleted turns out to be infinite cyclic, generated by the orange loop (or any other loop winding once around the hole). This way, the fundamental group detects the hole.

In more recent applications, the influence has also been reversed to motivate geometric constructions by a group-theoretical background. In a similar vein, geometric group theory employs geometric concepts, for example in the study of hyperbolic groups. Further branches crucially applying groups include algebraic geometry and number theory.

In addition to the above theoretical applications, many practical applications of groups exist. Cryptography relies on the combination of the abstract group theory approach together with algorithmical knowledge obtained in computational group theory, in particular when implemented for finite groups. Applications of group theory are not restricted to mathematics; sciences such as physics, chemistry and computer science benefit from the concept.

Numbers

Many number systems, such as the integers and the rationals enjoy a naturally given group structure. In some cases, such as with the rationals, both addition and multiplication operations give rise to group structures. Such number systems are predecessors to more general algebraic structures known as rings and fields. Further abstract algebraic concepts such as modules, vector spaces and algebras also form groups.

Integers

The group of integers Z under addition, denoted $(Z, +)$, has been described above. The integers, with the operation of multiplication instead of addition, $(Z, \cdot)$ do *not* form a group. The closure, associativity and identity axioms are satisfied, but inverses do not exist: for example, $a = 2$ is an integer, but the only solution to the equation $a \cdot b = 1$ in this case is $b = 1/2$, which is a rational number, but not an integer. Hence not every element of Z has a (multiplicative) inverse.

Rationals

The desire for the existence of multiplicative inverses suggests considering fractions

$$\frac{a}{b}.$$

Fractions of integers (with b nonzero) are known as rational numbers. The set of all such fractions is commonly denoted Q. There is still a minor obstacle for $(Q, \cdot)$, the rationals with multiplication, being a group: because the rational number 0 does not have a multiplicative inverse (i.e., there is no x such that $x \cdot 0 = 1$), $(Q, \cdot)$ is still not a group.

However, the set of all *nonzero* rational numbers $Q \setminus \{0\} = \{q \in Q \mid q \neq 0\}$ does form an abelian group under multiplication, denoted $(Q \setminus \{0\}, \cdot)$. Associativity and identity element axioms follow from the properties of integers. The closure requirement still holds true after removing zero, because the product of two nonzero rationals is never zero. Finally, the inverse of a/b is b/a, therefore the axiom of the inverse element is satisfied.

The rational numbers (including 0) also form a group under addition. Intertwining addition and multiplication operations yields more complicated structures called rings and—if division is possible, such as in Q—fields, which occupy a central position in abstract algebra. Group theoretic arguments therefore underlie parts of the theory of those entities.

Modular Arithmetic

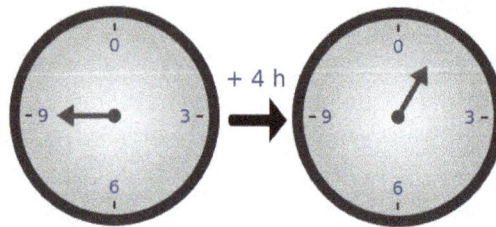

The hours on a clock form a group that uses addition modulo 12. Here 9 + 4 = 1.

In modular arithmetic, two integers are added and then the sum is divided by a positive integer called the *modulus*. The result of modular addition is the remainder of that division. For any modulus, n, the set of integers from 0 to $n - 1$ forms a group under modular addition: the inverse of any element a is $n - a$, and 0 is the identity element. This is familiar from the addition of hours on the face of a clock: if the hour hand is on 9 and is advanced 4 hours, it ends up on 1, as shown at the right. This is expressed by saying that 9 + 4 equals 1 "modulo 12" or, in symbols,

$9 + 4 \equiv 1$ modulo 12.

The group of integers modulo n is written Z_n or Z/nZ.

For any prime number p, there is also the multiplicative group of integers modulo p. Its elements are the integers 1 to $p - 1$. The group operation is multiplication modulo p. That is, the usual product is divided by p and the remainder of this division is the result of modular multiplication. For example, if $p = 5$, there are four group elements 1, 2, 3, 4. In this group, $4 \cdot 4 = 1$, because the usual product 16 is equivalent to 1, which divided by 5 yields a remainder of 1. for 5 divides $16 - 1 = 15$, denoted

$16 \equiv 1 \pmod 5$.

The primality of p ensures that the product of two integers neither of which is divisible by p is not divisible by p either, hence the indicated set of classes is closed under multiplication. The identity element is 1, as usual for a multiplicative group, and the associativity follows from the corresponding property of integers. Finally, the inverse element axiom requires that given an integer a not divisible by p, there exists an integer b such that

$a \cdot b \equiv 1 \pmod p$, i.e., p divides the difference $a \cdot b - 1$.

The inverse b can be found by using Bézout's identity and the fact that the greatest common divisor $\gcd(a, p)$ equals 1. In the case $p = 5$ above, the inverse of 4 is 4, and the inverse of 3 is 2, as 3

$\cdot\ 2 = 6 \equiv 1 \pmod 5$. Hence all group axioms are fulfilled. Actually, this example is similar to $(Q \setminus \{0\}, \cdot)$ above: it consists of exactly those elements in Z/pZ that have a multiplicative inverse. These groups are denoted $F_p^\times$. They are crucial to public-key cryptography.

Cyclic Groups

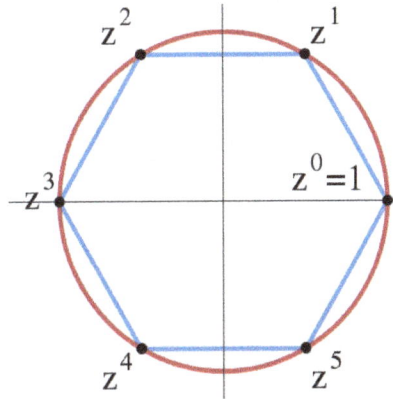

The 6th complex roots of unity form a cyclic group. z is a primitive element, but z^2 is not, because the odd powers of z are not a power of z^2.

A *cyclic group* is a group all of whose elements are powers of a particular element a. In multiplicative notation, the elements of the group are:

$$\dots, a^{-3}, a^{-2}, a^{-1}, a^0 = e, a, a^2, a^3, \dots,$$

where a^2 means $a \cdot a$, and a^{-3} stands for $a^{-1} \cdot a^{-1} \cdot a^{-1} = (a \cdot a \cdot a)^{-1}$ etc. Such an element a is called a generator or a primitive element of the group. In additive notation, the requirement for an element to be primitive is that each element of the group can be written as

$$\dots, -a-a, -a, 0, a, a+a, \dots$$

In the groups Z/nZ introduced above, the element 1 is primitive, so these groups are cyclic. Indeed, each element is expressible as a sum all of whose terms are 1. Any cyclic group with n elements is isomorphic to this group. A second example for cyclic groups is the group of n-th complex roots of unity, given by complex numbers z satisfying $z^n = 1$. These numbers can be visualized as the vertices on a regular n-gon, as shown in blue at the right for $n = 6$. The group operation is multiplication of complex numbers. In the picture, multiplying with z corresponds to a counter-clockwise rotation by 60°. Using some field theory, the group $F_p^\times$ can be shown to be cyclic: for example, if $p = 5$, 3 is a generator since $3^1 = 3$, $3^2 = 9 \equiv 4$, $3^3 \equiv 2$, and $3^4 \equiv 1$.

Some cyclic groups have an infinite number of elements. In these groups, for every non-zero element a, all the powers of a are distinct; despite the name "cyclic group", the powers of the elements do not cycle. An infinite cyclic group is isomorphic to $(Z, +)$, the group of integers under addition introduced above. As these two prototypes are both abelian, so is any cyclic group.

The study of finitely generated abelian groups is quite mature, including the fundamental theorem of finitely generated abelian groups; and reflecting this state of affairs, many group-related notions, such as center and commutator, describe the extent to which a given group is not abelian.

Symmetry Groups

Symmetry groups are groups consisting of symmetries of given mathematical objects—be they of geometric nature, such as the introductory symmetry group of the square, or of algebraic nature, such as polynomial equations and their solutions. Conceptually, group theory can be thought of as the study of symmetry. Symmetries in mathematics greatly simplify the study of geometrical or analytical objects. A group is said to act on another mathematical object X if every group element performs some operation on X compatibly to the group law. In the rightmost example below, an element of order 7 of the (2,3,7) triangle group acts on the tiling by permuting the highlighted warped triangles (and the other ones, too). By a group action, the group pattern is connected to the structure of the object being acted on.

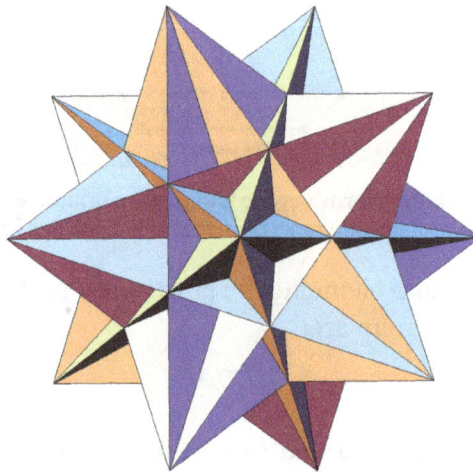

Rotations and reflections form the symmetry group of a great icosahedron.

In chemical fields, such as crystallography, space groups and point groups describe molecular symmetries and crystal symmetries. These symmetries underlie the chemical and physical behavior of these systems, and group theory enables simplification of quantum mechanical analysis of these properties. For example, group theory is used to show that optical transitions between certain quantum levels cannot occur simply because of the symmetry of the states involved.

Not only are groups useful to assess the implications of symmetries in molecules, but surprisingly they also predict that molecules sometimes can change symmetry. The Jahn-Teller effect is a distortion of a molecule of high symmetry when it adopts a particular ground state of lower symmetry from a set of possible ground states that are related to each other by the symmetry operations of the molecule.

Likewise, group theory helps predict the changes in physical properties that occur when a material undergoes a phase transition, for example, from a cubic to a tetrahedral crystalline form. An example is ferroelectric materials, where the change from a paraelectric to a ferroelectric state occurs at the Curie temperature and is related to a change from the high-symmetry paraelectric state to the lower symmetry ferroelectric state, accompanied by a so-called soft phonon mode, a vibrational lattice mode that goes to zero frequency at the transition.

Such spontaneous symmetry breaking has found further application in elementary particle physics, where its occurrence is related to the appearance of Goldstone bosons.

Buckminsterfuller-ene displays icosahedral symmetry, though the double bonds reduce this to pyritohedral symmetry.	Ammonia, NH_3. Its symmetry group is of order 6, generated by a 120° rotation and a reflection.	Cubane C_8H_8 features octahedral symmetry.	Hexaaquacopper(II) complex ion, $[Cu(OH_2)_6]^{2+}$. Compared to a perfectly symmetrical shape, the molecule is vertically dilated by about 22% (Jahn-Teller effect).	The (2,3,7) triangle group, a hyperbolic group, acts on this tiling of the hyperbolic plane.

Finite symmetry groups such as the Mathieu groups are used in coding theory, which is in turn applied in error correction of transmitted data, and in CD players. Another application is differential Galois theory, which characterizes functions having antiderivatives of a prescribed form, giving group-theoretic criteria for when solutions of certain differential equations are well-behaved. Geometric properties that remain stable under group actions are investigated in (geometric) invariant theory.

General Linear Group and Representation Theory

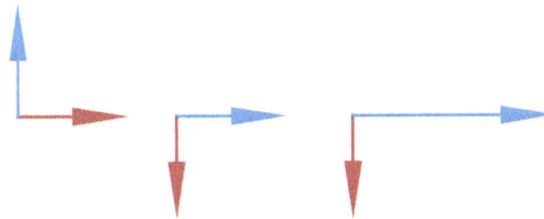

Two vectors (the left illustration) multiplied by matrices (the middle and right illustrations). The middle illustration represents a clockwise rotation by 90°, while the right-most one stretches the x-coordinate by factor 2.

Matrix groups consist of matrices together with matrix multiplication. The *general linear group* $GL(n, R)$ consists of all invertible n-by-n matrices with real entries. Its subgroups are referred to as *matrix groups* or *linear groups*. The dihedral group example mentioned above can be viewed as a (very small) matrix group. Another important matrix group is the special orthogonal group $SO(n)$. It describes all possible rotations in n dimensions. Via Euler angles, rotation matrices are used in computer graphics.

Representation theory is both an application of the group concept and important for a deeper understanding of groups. It studies the group by its group actions on other spaces. A broad class of group representations are linear representations, i.e., the group is acting on a vector space, such as the three-dimensional Euclidean space R^3. A representation of G on an n-dimensional real vector space is simply a group homomorphism

$$\rho: G \rightarrow GL(n, R)$$

from the group to the general linear group. This way, the group operation, which may be abstractly given, translates to the multiplication of matrices making it accessible to explicit computations.

Given a group action, this gives further means to study the object being acted on. On the other hand, it also yields information about the group. Group representations are an organizing principle in the theory of finite groups, Lie groups, algebraic groups and topological groups, especially (locally) compact groups.

Galois Groups

Galois groups were developed to help solve polynomial equations by capturing their symmetry features. For example, the solutions of the quadratic equation $ax^2 + bx + c = 0$ are given by

$$x = \frac{-b \pm \sqrt{b^2 - 4ac}}{2a}.$$

Exchanging "+" and "–" in the expression, i.e., permuting the two solutions of the equation can be viewed as a (very simple) group operation. Similar formulae are known for cubic and quartic equations, but do *not* exist in general for degree 5 and higher. Abstract properties of Galois groups associated with polynomials (in particular their solvability) give a criterion for polynomials that have all their solutions expressible by radicals, i.e., solutions expressible using solely addition, multiplication, and roots similar to the formula above.

The problem can be dealt with by shifting to field theory and considering the splitting field of a polynomial. Modern Galois theory generalizes the above type of Galois groups to field extensions and establishes—via the fundamental theorem of Galois theory—a precise relationship between fields and groups, underlining once again the ubiquity of groups in mathematics.

Finite Groups

A group is called *finite* if it has a finite number of elements. The number of elements is called the order of the group. An important class is the *symmetric groups* S_N, the groups of permutations of N letters. For example, the symmetric group on 3 letters S_3 is the group consisting of all possible orderings of the three letters *ABC*, i.e., contains the elements *ABC, ACB, BAC, BCA, CAB, CBA*, in total 6 (factorial of 3) elements. This class is fundamental insofar as any finite group can be expressed as a subgroup of a symmetric group S_N for a suitable integer N, according to Cayley's theorem. Parallel to the group of symmetries of the square above, S_3 can also be interpreted as the group of symmetries of an equilateral triangle.

The order of an element a in a group G is the least positive integer n such that $a^n = e$, where a^n represents

$$\underbrace{a \cdots a}_{n \text{ factors}},$$

i.e., application of the operation • to n copies of a. (If • represents multiplication, then a^n corresponds to the nth power of a.) In infinite groups, such an n may not exist, in which case the order of a is said to be infinity. The order of an element equals the order of the cyclic subgroup generated by this element.

More sophisticated counting techniques, for example counting cosets, yield more precise statements about finite groups: Lagrange's Theorem states that for a finite group G the order of any finite subgroup H divides the order of G. The Sylow theorems give a partial converse.

The dihedral group (discussed above) is a finite group of order 8. The order of r_1 is 4, as is the order of the subgroup R it generates. The order of the reflection elements f_v etc. is 2. Both orders divide 8, as predicted by Lagrange's theorem. The groups $F_p^{\times}$ above have order $p - 1$.

Classification of Finite Simple Groups

Mathematicians often strive for a complete classification of a mathematical notion. In the context of finite groups, this aim leads to difficult mathematics. According to Lagrange's theorem, finite groups of order p, a prime number, are necessarily cyclic (abelian) groups Z_p. Groups of order p^2 can also be shown to be abelian, a statement which does not generalize to order p^3, as the non-abelian group D_4 of order $8 = 2^3$ above shows. Computer algebra systems can be used to list small groups, but there is no classification of all finite groups. An intermediate step is the classification of finite simple groups. A nontrivial group is called *simple* if its only normal subgroups are the trivial group and the group itself. The Jordan–Hölder theorem exhibits finite simple groups as the building blocks for all finite groups. Listing all finite simple groups was a major achievement in contemporary group theory. 1998 Fields Medal winner Richard Borcherds succeeded in proving the monstrous moonshine conjectures, a surprising and deep relation between the largest finite simple sporadic group—the "monster group"—and certain modular functions, a piece of classical complex analysis, and string theory, a theory supposed to unify the description of many physical phenomena.

Groups with Additional Structure

Many groups are simultaneously groups and examples of other mathematical structures. In the language of category theory, they are group objects in a category, meaning that they are objects (that is, examples of another mathematical structure) which come with transformations (called morphisms) that mimic the group axioms. For example, every group (as defined above) is also a set, so a group is a group object in the category of sets.

Topological Groups

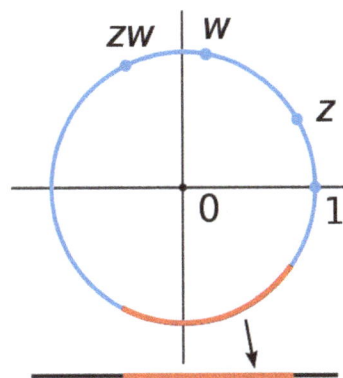

The unit circle in the complex plane under complex multiplication is a Lie group and, therefore, a topological group. It is topological since complex multiplication and division are continuous. It is a manifold and thus a Lie group, because every small piece, such as the red arc in the figure, looks like a part of the real line (shown at the bottom).

Some topological spaces may be endowed with a group law. In order for the group law and the topology to interweave well, the group operations must be continuous functions, that is, $g \cdot h$, and g^{-1} must not vary wildly if g and h vary only little. Such groups are called *topological groups,* and they are the group objects in the category of topological spaces. The most basic examples are the reals R under addition, $(R \setminus \{0\}, \cdot)$, and similarly with any other topological field such as the complex numbers or *p*-adic numbers. All of these groups are locally compact, so they have Haar measures and can be studied via harmonic analysis. The former offer an abstract formalism of invariant integrals. Invariance means, in the case of real numbers for example:

$$\int f(x)dx = \int f(x+c)dx$$

for any constant *c*. Matrix groups over these fields fall under this regime, as do adele rings and adelic algebraic groups, which are basic to number theory. Galois groups of infinite field extensions such as the absolute Galois group can also be equipped with a topology, the so-called Krull topology, which in turn is central to generalize the above sketched connection of fields and groups to infinite field extensions. An advanced generalization of this idea, adapted to the needs of algebraic geometry, is the étale fundamental group.

Lie Groups

Lie groups (in honor of Sophus Lie) are groups which also have a manifold structure, i.e., they are spaces looking locally like some Euclidean space of the appropriate dimension. Again, the additional structure, here the manifold structure, has to be compatible, i.e., the maps corresponding to multiplication and the inverse have to be smooth.

A standard example is the general linear group introduced above: it is an open subset of the space of all *n*-by-*n* matrices, because it is given by the inequality

$$\det (A) \neq 0,$$

where *A* denotes an *n*-by-*n* matrix.

Lie groups are of fundamental importance in modern physics: Noether's theorem links continuous symmetries to conserved quantities. Rotation, as well as translations in space and time are basic symmetries of the laws of mechanics. They can, for instance, be used to construct simple models—imposing, say, axial symmetry on a situation will typically lead to significant simplification in the equations one needs to solve to provide a physical description. Another example are the Lorentz transformations, which relate measurements of time and velocity of two observers in motion relative to each other. They can be deduced in a purely group-theoretical way, by expressing the transformations as a rotational symmetry of Minkowski space. The latter serves—in the absence of significant gravitation—as a model of space time in special relativity. The full symmetry group of Minkowski space, i.e., including translations, is known as the Poincaré group. By the above, it plays a pivotal role in special relativity and, by implication, for quantum field theories. Symmetries that vary with location are central to the modern description of physical interactions with the help of gauge theory.

Generalizations

Group-like structures					
	Totality[α]	**Associativity**	**Identity**	**Invertibility**	**Commutativity**
Semicategory	Unneeded	Required	Unneeded	Unneeded	Unneeded
Category	Unneeded	Required	Required	Unneeded	Unneeded
Groupoid	Unneeded	Required	Required	Required	Unneeded
Magma	Required	Unneeded	Unneeded	Unneeded	Unneeded
Quasigroup	Required	Unneeded	Unneeded	Required	Unneeded
Loop	Required	Unneeded	Required	Required	Unneeded
Semigroup	Required	Required	Unneeded	Unneeded	Unneeded
Monoid	Required	Required	Required	Unneeded	Unneeded
Group	Required	Required	Required	Required	Unneeded
Abelian group	Required	Required	Required	Required	Required

^α Closure, which is used in many sources, is an equivalent axiom to totality, though defined differently.

In abstract algebra, more general structures are defined by relaxing some of the axioms defining a group. For example, if the requirement that every element has an inverse is eliminated, the resulting algebraic structure is called a monoid. The natural numbers N (including 0) under addition form a monoid, as do the nonzero integers under multiplication (Z \ {0}, ·). There is a general method to formally add inverses to elements to any (abelian) monoid, much the same way as (Q \ {0}, ·) is derived from (Z \ {0}, ·), known as the Grothendieck group. Groupoids are similar to groups except that the composition $a \cdot b$ need not be defined for all a and b. They arise in the study of more complicated forms of symmetry, often in topological and analytical structures, such as the fundamental groupoid or stacks. Finally, it is possible to generalize any of these concepts by replacing the binary operation with an arbitrary n-ary one (i.e., an operation taking n arguments). With the proper generalization of the group axioms this gives rise to an n-ary group. The table gives a list of several structures generalizing groups.

Our aim is to look at groups and use it to the study of questions of the type:

1. How many different necklace configurations are possible if we use 6 beads of 3 different colors? Or, for that matter what if we use n beads of m different colors?

2. How many different necklace configurations are possible if we use 12 beads among which 3 are red, 5 are blue and 4 are green? And a generalization of this problem.

3. Counting the number of chemical compounds which can be derived by the substitution of a given set of radicals in a given molecular structure.

It can be easily observed that if we want to look at different color configurations of a necklace formed using 6 beads, we need to understand the symmetries of a hexagon. Such a study is achieved through what in literature is called groups. Once we have learnt a bit about groups, we study group action. This study helps us in defining an equivalence relation on the set of color configurations for a given necklace. And it turns out that the number of distinct color configurations is same as the number of equivalence classes.

Before coming to the definition and its properties, let us look at the properties of the sets $\mathbb{N}, \mathbb{Z}, \mathbb{Q}, \mathbb{R}$, and $\mathbb{C}$. We know that the sets $\mathbb{Z}, \mathbb{Q}, \mathbb{R}$ and $\mathbb{C}$ satisfy the following:

Binary Operation: for every a, $b \in \mathbb{Z}(\mathbb{Q}, \mathbb{R}, \mathbb{C}), a + b$, called the addition of a and b, is an element of $\mathbb{Z}(\mathbb{Q}, \mathbb{R}, \mathbb{C})$;

Addition is Associative: for every $a, b, c \in \mathbb{Z}(\mathbb{Q}, \mathbb{R}, \mathbb{C}), (a + b) + c = a + (b + c)$;

Additive Identity: the element zero, denoted 0, is an element of $\mathbb{Z}(\mathbb{Q}, \mathbb{R}, \mathbb{C})$ and has the property that for every $a \in \mathbb{Z}(\mathbb{Q}, \mathbb{R}, \mathbb{C}), a + 0 = a = 0 + a$;

Additive Inverse: For every element $a \in \mathbb{Z}(\mathbb{Q}, \mathbb{R}, \mathbb{C})$, there exists an element $-a \in \mathbb{Z}(\mathbb{Q}, \mathbb{R}, \mathbb{C})$ such that a + (−a) = 0 = −a + a;

Addition is Commutative: We also have a + b = b + a for every $a, b \in \mathbb{Z}(\mathbb{Q}, \mathbb{R}, \mathbb{C})$.

Now, let us look at the sets $\mathbb{Z}^* = \mathbb{Z} - \{0\}, \mathbb{Q}^* = \mathbb{Q} - \{0\}, \mathbb{R}^* = \mathbb{R} - \{0\}$ and $\mathbb{C}^* = \mathbb{C} - \{0\}$. As in the previous case, we see that similar statements hold true for the sets $\mathbb{Z}^*, \mathbb{Q}^*, \mathbb{R}^*$ and $\mathbb{C}^*$. Namely,

Binary Operation: for every $a, b \in \mathbb{Z}^*(\mathbb{Q}^*, \mathbb{R}^*, \mathbb{C}^*), a \bullet b$, called the multiplication of a and b, is an element of $\mathbb{Z}^*(\mathbb{Q}^*, \mathbb{R}^*, \mathbb{C}^*)$;

Multiplication is Associative: for every $a, b, c \in \mathbb{Z}^*(\mathbb{Q}^*, \mathbb{R}^*, \mathbb{C}^*), (a \bullet b) \bullet c = a \bullet (b \bullet c)$;

Multiplicative Identity: the element one, denoted 1, is an element of $\mathbb{Z}^*(\mathbb{Q}^*, \mathbb{R}^*, \mathbb{C}^*)$ and for all $a \in \mathbb{Z}^*(\mathbb{Q}^*, \mathbb{R}^*, \mathbb{C}^*), a \bullet 1 = a = 1 \bullet a$;

Multiplication is Commutative: One also has a • b = b • a for every $a, b \in \mathbb{Z}^*(\mathbb{Q}^*, \mathbb{R}^*, \mathbb{C}^*)$.

Observe that if we choose $a \in \mathbb{Z}^*$ with $a \neq 1, -1$ then there does not exist an element $b \in \mathbb{Z}^*$ such that a • b = 1 = b • a. Whereas, for the sets $\mathbb{Q}^*, \mathbb{R}^*$ and $\mathbb{C}^*$ one can always find a b such that a • b = 1 = b • a.

Based on the above examples, an abstract notion called groups is defined. Formally, one defines a group as follows.

Definition(1) (Group). A group G, usually denoted (G, *), is a non-empty set, together with a binary operation, say *, such that the elements of G satisfy the following:

1. for every a, b, c ∈ G, (a * b) * c = a * (b * c) (Associativity Property) holds in G;

2. there is an element e ∈ G such that a * e = a = e * a, for all a ∈ G (Existence of Identity);

3. for every element a ∈ G, there exists an element b ∈ G such that a*b = e = b*a (Existence of Inverse).

In addition, if the set G satisfies a * b = b * a, for every a, b ∈ G, then G is said to be an abelian (commutative) group.

Before proceeding with examples of groups that concerns us, we state a few basic results in group theory without proof. The readers are advised to prove it for themselves.

Remark: Let (G, *) be a group. Then the following hold:

1. The identity element of G is unique. Hence, the identity element is denoted by e.

2. For each fixed a ∈ G, the element b ∈ G such that a*b = e = b*a is also unique. Therefore, for each a ∈ G, the element b that satisfies a * b = e = b * a is denoted by a^{-1}.

3. Also, for each a ∈ G, $(a^{-1})^{-1}$ = a.

4. If a * b = a * c, for some a, b, c ∈ G then b = c. Similarly, if b * d = c * d, for some b, c, d ∈ G then b = c. That is, the cancelation laws in G.

5. For each a, b ∈ G, $(ab)^{-1}$ = $b^{-1}a^{-1}$.

6. By convention, we assume a^0 = e, for all a ∈ G.

7. For each a ∈ G, $(a^n)^{-1}$ = $(a^{-1})^n$, for all $n \in \mathbb{Z}$.

Example: Symmetric group on n letters: Let N denote the set {1, 2, . . . , n}. A function f : N →N is called a permutation on n elements if f is both one-to-one and onto. Let S_n = {f : N →N | f is one to one and onto}. That is, S_n is the set of all permutations of the set {1, 2, . . . , n}. Then the following can be verified:

1. Suppose f, g ∈ S_n. Then f : N →N and g : N →N are two one-to-one and onto functions. Therefore, one uses the composition of functions to define the composite function $f \circ g : N \to N$ by $(f \circ g)(x) = f(g(x))$. Then it can be easily verified that f ∘g is also one-to-one and onto. Hence f ∘ g ∈ S_n. That is, "composition of function", denoted ∘, defines a binary operation in S_n.

2. It is well known that the composition of functions is an associative operation and thus (f ∘ g) ∘ h = f ∘ (g ∘ h).

3. The function e : N →N defined by e(i) = i, for all i = 1, 2, . . . , n is the identity function. That is, check that f ∘ e = f = e ∘ f, for all f ∈ S_n.

4. Now let f ∈ S_n. As f : N →N is a one-to-one and onto function, f^{-1} : N →N defined by $f^{-1}(i)$ = j, whenever f (j) = i, for all i = 1, 2, . . . , n, is a well defined function and is also one-to-one and onto. That is, for each f ∈ S_n, f^{-1} ∈ S_n and f ∘ f^{-1} = e = f^{-1} ∘ f.

Thus (S_n, ∘) is a group. This group is called the Symmetric/Permutation group on n letters. If σ ∈ S_n then one represents this by writing $\sigma = \begin{pmatrix} 1 & 2 & \cdots & n \\ \sigma(1) & \sigma(2) & \cdots & \sigma(n) \end{pmatrix}$. This representation of an element of S_n is called a two row notation. Observe that as σ is one-to-one and onto function from N to N, it can be checked that N = {σ(1), σ(2), . . . , σ(n)}. Hence, there are n choices for σ(1), n − 1 choices for σ(2) (all elements of N except σ(1)) and so on. Thus, the total number of elements in S_n is n! = n(n − 1) • • • 2 • 1.

Before discussing other examples, let us try to understand the group S_n. As seen above, any element σ ∈ S_n can be represented using a two-row notation. There is another notation for permutations that is often very useful. This notation is called the cycle notation. Let us try to understand this notation.

Definition: Let $\sigma \in S_n$ and let $S = \{i_1, i_2, \ldots, i_k\} \subseteq \{1, 2, \ldots, n\}$ be distinct. If σ satisfies

$$\sigma(i_\ell) = i_{\ell+1}, \text{for all } \ell = 1, 2, \ldots, k-1, \sigma(i_k) = i_1 \text{ and } \sigma(r) = r \text{ for } r \notin S$$

then σ is called a k-cycle and is denoted by $\sigma = (i_1, i_2, \ldots i_k)$ or $(i_2, i_3, \ldots, i_k, i_1)$ and so on.

Example: 1. The permutation $\sigma = \begin{pmatrix} 1 & 2 & 3 & 4 & 5 \\ 2 & 3 & 4 & 1 & 5 \end{pmatrix}$ in cycle notation can be written

as (1234), (2341), (3412), or (4123) as $\sigma(1) = 2$, $\sigma(2) = 3$, $\sigma(3) = 4$, $\sigma(4) = 1$ and $\sigma(5) = 5$.

2. The permutation $\begin{pmatrix} 1 & 2 & 3 & 4 & 5 & 6 \\ 2 & 3 & 1 & 4 & 6 & 5 \end{pmatrix}$ in cycle notation equals (123)(65) as $\sigma(1) = 2$, $\sigma(2) = 3$, $\sigma(3)$ = 1, $\sigma(4) = 4$, $\sigma(5) = 6$ and $\sigma(6) = 5$. That is, this element is formed with the help of two cycles (123) and (56).

3. Consider two permutations $\sigma = (143)(27)$ and $\tau = (1357)(246)$. Then, there composition, denoted $\sigma \circ \tau$, is obtained as follows:

$(\sigma \circ \tau)(1) = \sigma(\tau(1)) = \sigma(3) = 1$, $(\sigma \circ \tau)(2) = \sigma(\tau(2)) = \sigma(4) = 3$, $(\sigma \circ \tau)(3) = \sigma(\tau(3)) = \sigma(5) = 5$, $(\sigma \circ \tau)(4) = \sigma(\tau(4)) = \sigma(6) = 6$, $(\sigma \circ \tau)(5) = 2$, $(\sigma \circ \tau)(6) = 7$ and $(\sigma \circ \tau)(7) = 4$.

Hence

$$\sigma \circ \tau = (143)(27)(1357)(246) = \begin{pmatrix} 1 & 2 & 3 & 4 & 5 & 6 & 7 \\ 1 & 3 & 5 & 6 & 2 & 7 & 4 \end{pmatrix} = (235)(467).$$

4. Similarly, verify that (1456)(152) = (16)(245).

Definition: Two cycles $\sigma = (i_1, i_2, \ldots, i_t)$ and $\tau = (j_1, j_2, \ldots, j_s)$ are said to be disjoint if

$$\{i_1, i_2, \ldots, i_t\} \cap \{j_1, j_2, \ldots, j_s\} = \emptyset.$$

The proof of the following theorem can be obtained from any standard book on abstract algebra.

Theorem: Let $\sigma \in S_n$. Then σ can be written as a product of disjoint cycles.

Remark: Observe that the representation of a permutation as a product of disjoint cycles, none of which is the identity, is unique up to the order of the disjoint cycles. The representation of an element $\sigma \in S_n$ as product of disjoint cycles is called the cyclic decomposition of σ.

Symmetries of regular n-gons in plane.

(a) Suppose a unit square is placed at the coordinates (0, 0, 0), (1, 0, 0), (0, 1, 0) and (1, 1, 0). Our aim is to move the square in space such that the position of the vertices may change but they are still placed at the above mentioned coordinates. The question arises, what are the possible ways in which this can be done? It can be easily verified that the possible configurations are as follows (see figure a below):

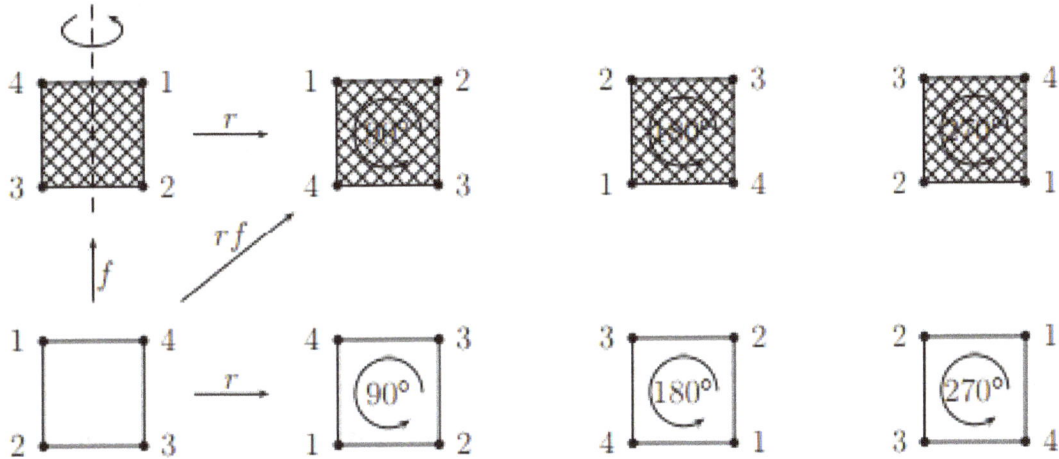

(a) Symmetries of a square.

Let r denote the counter-clockwise rotation of the square by 90° and f denote the flipping of the square along the vertical axis passing through the midpoint of opposite horizontal edges. Then note that the possible configurations corre- spond to the set

$$G = \left\{ e, r, r^2, r^3, f, rf, r^2 f, r^3 f \right\} \text{ with relations } r^4 = e = f^2 \text{ and } fr^3 = rf.$$

Using the above equation, observe that $(rf)^2 = (rf)(rf) = r(fr)f = r(r^3f)f = r^4f^2 = e$. Similarly, it can be checked that $(r^2f)^2 = (r^3f)^2 = e$. That is, all the terms f, rf, r²f and r³f are flips. The group G is generally denoted by D_4 and is called the Dihedral group or the symmetries of a square. This group can also be represented as follows:

{e, (1234), (13)(24), (1432), (14)(23), (24), (12)(34), (13)}.

Exercise: Relate the two representations of the group D_4.

(b) In the same way, one can define the symmetries of an equilateral triangle (see figure b). This group is denoted by D_3 and is represented as

$$D_3 = \left\{ e, r, r^2, f, rf, r^2 f \right\} \text{ with relations } r^3 = e = f^2 \text{ and } fr^2 = rf,$$

where r is a counter-clockwise rotation by $120° = \dfrac{2\pi}{3}$ and f is a flip. Using the following figure, one can check that the group D_3 can also be represented by

D_3 = {e, (ABC), (ACB), (BC), (CA), (AB)}.

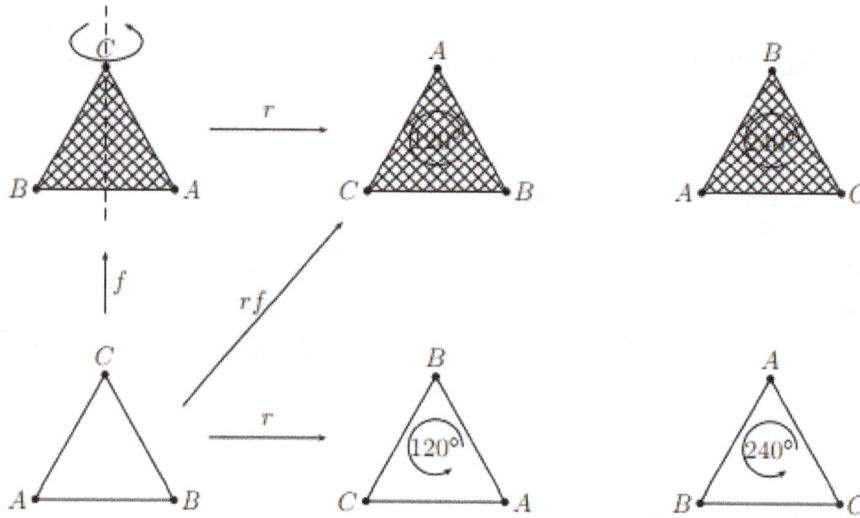

(b) Symmetries of an Equilateral Triangle.

(c) For a regular pentagon, it can be verified that the group of symmetries of a regular pentagon is given by G = {e, r, r², r³, r⁴, f, rf, r²f, r³f, r⁴f } with $r^5 = 2 = f^2$ and $rf = fr^4$, where r denotes a counter-clockwise rotation through an angle of $72° = \dfrac{2\pi}{5}$ and f is a flip along a line that passes through a vertex and the midpoint of the opposite edge. Or equivalently, if we label the vertices of a regular pentagon, counter-clockwise, with the numbers 1, 2, 3, 4 and 5 then

G = {e, (1, 2, 3, 4, 5), (1, 3, 5, 2, 4), (1, 4, 2, 5, 3), (1, 5, 4, 3, 2), (2, 5)(3, 4),

(1, 3)(4, 5), (1, 5)(2, 4), (1, 2)(3, 5), (1, 4)(2, 3)}.

(d) In general, one can define symmetries of a regular n-gon. This group is denoted by D_n, has 2n elements and is represented as

$$\left\{e,\ r, r^2,\ \dots,\ r^{n-1}, f,\ rf,\ \dots,\ r^{n-1}f\right\}\ with\ r^n = e = f^2\ and\ fr^{n-1} = rf.$$

Here the symbol r stands for a counter-clockwise rotation through an angle of $\dfrac{2\pi}{n}$ and f stands for a vertical flip.

2. Symmetries of regular platonic solids.

(a) Recall from geometry that a tetrahedron is a 3-dimensional regular object that is composed of 4-equilateral triangles such that any three triangles meet at a vertex (see figure a). Observe that a tetrahedron has 6 edges, 4 vertices and 4 faces. If we denote the vertices of the tetrahedron with numbers 1, 2, 3 and 4, then the symmetries of the tetrahedron can be represented with the help of the group,

G = {e, (234), (243), (124), (142), (123), (132), (134), (143), (12)(34), (13)(24), (14)(23)},

where, for distinct numbers i, j, k and ℓ, the element (ijk) is formed by a rotation of 120° along the line that passes through the vertex ℓ and the centroid of the equilateral triangle with vertices i, j and k. Similarly, the group element (ij)(kℓ) is formed by a rotation of 180° along the line that passes through mid-points of the edges (ij) and (kℓ).

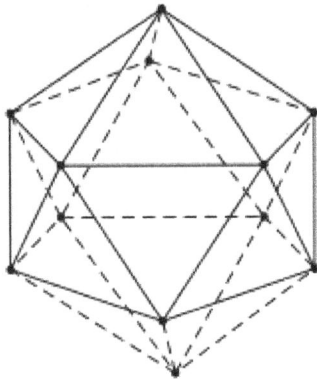

An Icosahedron

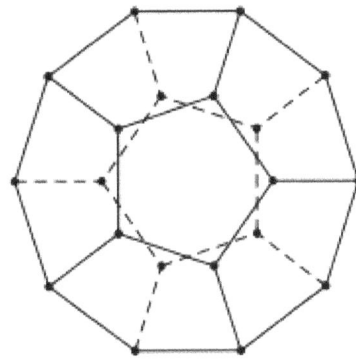

A Dodecahedron

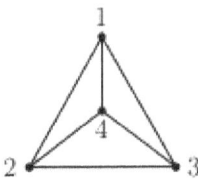

A Tetrahedron

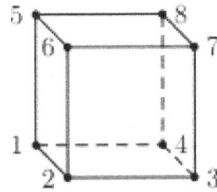

A Cube

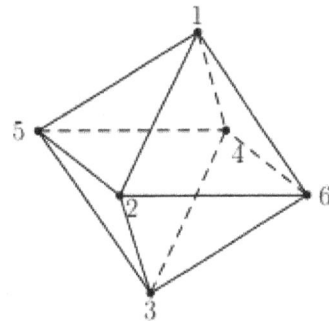

An Octahedron

(c) Regular Platonic solids.

(b) Consider the Cube and the Octahedron given in figure c. It can be checked that the group of symmetries of the two figures has 24 elements. We give the group elements for the symmetries of the cube, when the vertices of the cube are labeled. The readers are required to compute the group elements for the symmetries of the octahedron. For the cube (see figure c), the group elements are

i). e, the identity element;

ii). $3 \times 3 = 9$ elements that are obtained by rotations along lines that pass through the center of opposite faces (3 pairs of opposite faces and each face is a square: corresponds to a rotation of 90°). In terms of the vertices of the cube, the group elements are

(1234)(5678), (13)(24)(57)(68), (1432)(5876), (1265)(3784), (16)(25)(38)(47), (1562)(3487), (1485)(2376), (18)(45)(27)(36), (1584)(2673).

iii). $2 \times 4 = 8$ elements that are obtained by rotations along lines that pass through opposite vertices (4 pairs of opposite vertices and each vertex is incident with 3 edges: correspond to a rotation of 120°). The group elements in terms of the vertices of the cube are

(254)(368), (245)(386), (163)(457), (136)(475), (275)(138), (257)(183), (168)(274), (186)(247).

iv). $1 \times 6 = 6$ elements that are obtained by rotations along lines that pass through the midpoint of opposite edges (6 pairs of opposite edges: corresponds to a rotation of 180°). The corresponding elements in terms of the vertices of the cube are

(14)(67)(28)(35), (23)(58)(17)(46), (15)(37)(28)(64), (26)(48)(17)(35), (12)(78)(35)(46), (34)(56)(17)(28).

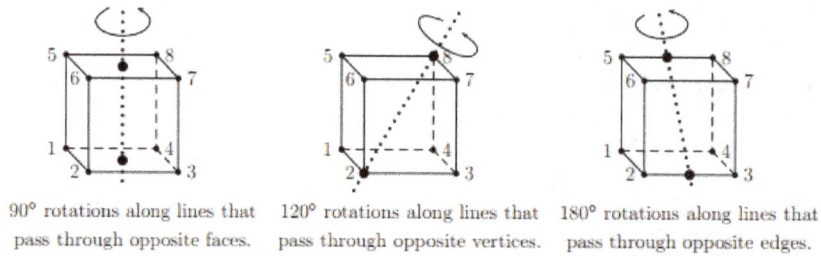

90° rotations along lines that pass through opposite faces. 120° rotations along lines that pass through opposite vertices. 180° rotations along lines that pass through opposite edges.

(d) Understanding the group of symmetries of a cube.

(c) Consider now the icosahedron and the dodecahedron (see figure c). Note that the icosahedron has 12 vertices, 20 faces and 30 edges and the dodecahedron has 20 vertices, 12 faces and 30 edges. It can be checked that the group of symmetries of the two figures has 60 elements. We give the idea of the group elements for the symmetries of the icosahedron. The readers are required to compute the group elements for the symmetries of the dodecahedron. For the icosahedron, one has

i).   e, the identity element;

ii). $2 \times 10 = 20$ elements that are obtained by rotations along lines that pass through the center of opposite faces (10 pairs of opposite faces and each face is an equilateral triangle: corresponds to a rotation of 120°);

iii). $6 \times 4 = 24$ elements that are obtained by rotations along lines that pass through opposite vertices (6 pairs of opposite vertices and each vertex is incident with 5 edges: corresponds to a rotation of 72°);

iv). $1 \times 15 = 15$ elements that are obtained by rotations along lines that pass through the midpoint of opposite edges (15 pairs of opposite edges: corresponds to a rotation of 180°).

Group Homomorphism

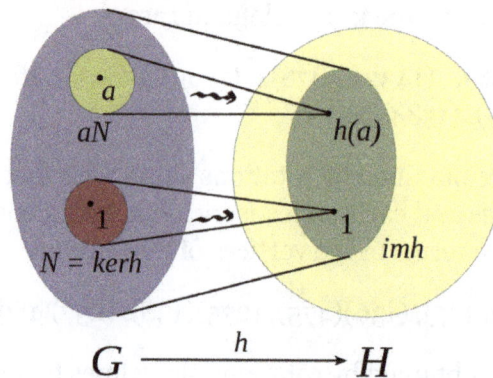

Image of a group homomorphism (h) from G (left) to H (right). The smaller oval inside H is the image of h. N is the kernel of h and aN is a coset of N.

In mathematics, given two groups, $(G, *)$ and $(H, \cdot)$, a group homomorphism from $(G, *)$ to $(H, \cdot)$ is a function $h : G \to H$ such that for all u and v in G it holds that

$$h(u * v) = h(u) \cdot h(v)$$

where the group operation on the left hand side of the equation is that of G and on the right hand side that of H.

From this property, one can deduce that h maps the identity element e_G of G to the identity element e_H of H, and it also maps inverses to inverses in the sense that

$$h\left(u^{-1}\right) = h(u)^{-1}.$$

Hence one can say that h "is compatible with the group structure".

Older notations for the homomorphism $h(x)$ may be x_h, though this may be confused as an index or a general subscript. A more recent trend is to write group homomorphisms on the right of their arguments, omitting brackets, so that $h(x)$ becomes simply $x\,h$. This approach is especially prevalent in areas of group theory where automata play a role, since it accords better with the convention that automata read words from left to right.

In areas of mathematics where one considers groups endowed with additional structure, a *homomorphism* sometimes means a map which respects not only the group structure (as above) but also the extra structure. For example, a homomorphism of topological groups is often required to be continuous.

Intuition

The purpose of defining a group homomorphism is to create functions that preserve the algebraic structure. An equivalent definition of group homomorphism is: The function $h : G \to H$ is a group homomorphism if whenever $a * b = c$ we have $h(a) \cdot h(b) = h(c)$. In other words, the group H in some sense has a similar algebraic structure as G and the homomorphism h preserves that.

Types of Group Homomorphism

Monomorphism

A group homomorphism that is injective (or, one-to-one); i.e., preserves distinctness.

Epimorphism

A group homomorphism that is surjective (or, onto); i.e., reaches every point in the codomain.

Isomorphism

A group homomorphism that is bijective; i.e., injective and surjective. Its inverse is also a group homomorphism. In this case, the groups G and H are called isomorphic; they differ only in the notation of their elements and are identical for all practical purposes.

Endomorphism

A homomorphism, h: G → G; the domain and codomain are the same. Also called an endomorphism of G.

Automorphism

An endomorphism that is bijective, and hence an isomorphism. The set of all automor-phisms of a group G, with functional composition as operation, forms itself a group, the automorphism group of G. It is denoted by Aut(G). As an example, the automorphism group of (Z, +) contains only two elements, the identity transformation and multiplication with −1; it is isomorphic to Z/2Z.

Image and Kernel

We define the *kernel of h* to be the set of elements in *G* which are mapped to the identity in *H*

$$\ker(h) \equiv \left\{ u \in G : h(u) = e_H \right\}.$$

and the *image of h* to be

$$\operatorname{im}(h) \equiv h(G) \equiv \left\{ h(u) : u \in G \right\}.$$

The kernel and image of a homomorphism can be interpreted as measuring how close it is to being an isomorphism. The first isomorphism theorem states that the image of a group homomorphism, $h(G)$ is isomorphic to the quotient group $G/\ker h$.

The kernel of h is a normal subgroup of *G* and the image of h is a subgroup of *H*:

$$\left(g^{-1} o u o g \right) = h(g)^{-1} \cdot h(u) \cdot h(g)$$
$$= h(g)^{-1} \cdot e_H \cdot h(g)$$
$$= h(g)^{-1} \cdot h(g) = e_H.$$

If and only if ker(*h*) = {e_G}, the homomorphism, *h*, is a *group monomorphism*; i.e., *h* is injective (one-to-one). Injection directly gives that there is a unique element in the kernel, and a unique element in the kernel gives injection:

$$h(g_1) = h(g_2)$$
$$\Leftrightarrow h(g_1) \cdot h(g_2)^{-1} = e_H$$
$$\Leftrightarrow h\left(g_1 o g_2^{-1} \right) = e_H, \ker(h) = \{e_G\}$$
$$\Rightarrow g_1 o g_2^{-1} = e_G$$
$$\Leftrightarrow g_1 = g_2$$

Examples

- Consider the cyclic group Z/3Z = {0, 1, 2} and the group of integers Z with addition. The map $h : Z \rightarrow Z/3Z$ with $h(u) = u$ mod 3 is a group homomorphism. It is surjective and its kernel consists of all integers which are divisible by 3.

- Consider the group

$$G \equiv \left\{ \begin{pmatrix} a & b \\ 0 & 1 \end{pmatrix} \middle| a > 0, b \in \mathbf{R} \right\}$$

For any complex number u the function $f_u : G \to C$ defined by:

$$\begin{pmatrix} a & b \\ 0 & 1 \end{pmatrix} \mapsto a^u$$

is a group homomorphism.

- Consider multiplicative group of positive real numbers $(\mathbf{R}^+, \cdot)$ for any complex number u the function $f_u : \mathbf{R}^+ \to C$ defined by:

$$f_u(a) = a^u$$

is a group homomorphism.

- The exponential map yields a group homomorphism from the group of real numbers R with addition to the group of non-zero real numbers R* with multiplication. The kernel is {0} and the image consists of the positive real numbers.

- The exponential map also yields a group homomorphism from the group of complex numbers C with addition to the group of non-zero complex numbers C* with multiplication. This map is surjective and has the kernel $\{2\pi k i : k \in Z\}$, as can be seen from Euler's formula. Fields like R and C that have homomorphisms from their additive group to their multiplicative group are thus called exponential fields.

The Category of Groups

If $h : G \to H$ and $k : H \to K$ are group homomorphisms, then so is $k \circ h : G \to K$. This shows that the class of all groups, together with group homomorphisms as morphisms, forms a category.

Homomorphisms of Abelian Groups

If G and H are abelian (i.e., commutative) groups, then the set Hom(G, H) of all group homomorphisms from G to H is itself an abelian group: the sum $h + k$ of two homomorphisms is defined by

$(h + k)(u) = h(u) + k(u)$ for all u in G.

The commutativity of H is needed to prove that $h + k$ is again a group homomorphism.

The addition of homomorphisms is compatible with the composition of homomorphisms in the following sense: if f is in Hom(K, G), h, k are elements of Hom(G, H), and g is in Hom(H, L), then

$(h + k) \circ f = (h \circ f) + (k \circ f)$ and $g \circ (h + k) = (g \circ h) + (g \circ k)$.

Since the composition is associative, this shows that the set End(G) of all endomorphisms of an abelian group forms a ring, the *endomorphism ring* of G. For example, the endomorphism ring of the abelian group consisting of the direct sum of m copies of Z/nZ is isomorphic to the ring of m-by-m matrices with entries in Z/nZ. The above compatibility also shows that the category of all abelian groups with group homomorphisms forms a preadditive category; the existence of direct sums and well-behaved kernels makes this category the prototypical example of an abelian category.

Subgroup

In group theory, a branch of mathematics, given a group G under a binary operation $*$, a subset H of G is called a subgroup of G if H also forms a group under the operation $*$. More precisely, H is a subgroup of G if the restriction of $*$ to $H \times H$ is a group operation on H. This is usually denoted $H \leq G$, read as "H is a subgroup of G".

The trivial subgroup of any group is the subgroup $\{e\}$ consisting of just the identity element.

A proper subgroup of a group G is a subgroup H which is a proper subset of G (i.e. $H \neq G$). This is usually represented notationally by $H < G$, read as "H is a proper subgroup of G". Some authors also exclude the trivial group from being proper (i.e. $\{e\} \neq H \neq G$).

If H is a subgroup of G, then G is sometimes called an overgroup of H.

The same definitions apply more generally when G is an arbitrary semigroup, but this article will only deal with subgroups of groups. The group G is sometimes denoted by the ordered pair $(G, *)$, usually to emphasize the operation $*$ when G carries multiple algebraic or other structures.

This article will write ab for $a * b$, as is usual.

Basic Properties of Subgroups

- A subset H of the group G is a subgroup of G if and only if it is nonempty and closed under products and inverses. (The closure conditions mean the following: whenever a and b are in H, then ab and a^{-1} are also in H. These two conditions can be combined into one equivalent condition: whenever a and b are in H, then ab^{-1} is also in H.) In the case that H is finite, then H is a subgroup if and only if H is closed under products. (In this case, every element a of H generates a finite cyclic subgroup of H, and the inverse of a is then $a^{-1} = a^{n-1}$, where n is the order of a.)

- The above condition can be stated in terms of a homomorphism; that is, H is a subgroup of a group G if and only if H is a subset of G and there is an inclusion homomorphism (i.e., i(a) = a for every a) from H to G.

- The identity of a subgroup is the identity of the group: if G is a group with identity e_G, and H is a subgroup of G with identity e_H, then $e_H = e_G$.

- The inverse of an element in a subgroup is the inverse of the element in the group: if H is a subgroup of a group G, and a and b are elements of H such that $ab = ba = e_H$, then $ab = ba = e_G$.

- The intersection of subgroups *A* and *B* is again a subgroup. The union of subgroups *A* and *B* is a subgroup if and only if either *A* or *B* contains the other, since for example 2 and 3 are in the union of 2Z and 3Z but their sum 5 is not. Another example is the union of the x-axis and the y-axis in the plane (with the addition operation); each of these objects is a subgroup but their union is not. This also serves as an example of two subgroups, whose intersection is precisely the identity.

- If *S* is a subset of *G*, then there exists a minimum subgroup containing *S*, which can be found by taking the intersection of all of subgroups containing *S*; it is denoted by <*S*> and is said to be the subgroup generated by *S*. An element of *G* is in <*S*> if and only if it is a finite product of elements of *S* and their inverses.

- Every element *a* of a group *G* generates the cyclic subgroup <*a*>. If <*a*> is isomorphic to Z/*n*Z for some positive integer *n*, then *n* is the smallest positive integer for which $a^n = e$, and *n* is called the *order* of *a*. If <*a*> is isomorphic to Z, then *a* is said to have *infinite order*.

- The subgroups of any given group form a complete lattice under inclusion, called the lattice of subgroups. (While the infimum here is the usual set-theoretic intersection, the supremum of a set of subgroups is the subgroup *generated by* the set-theoretic union of the subgroups, not the set-theoretic union itself.) If *e* is the identity of *G*, then the trivial group {*e*} is the minimum subgroup of *G*, while the maximum subgroup is the group *G* itself.

G is the group $\mathbb{Z}/8\mathbb{Z}$, the integers mod 8 under addition. The subgroup H contains only 0 and 4, and is isomorphic to $\mathbb{Z}/2\mathbb{Z}$. There are four left cosets of H: H itself, 1+H, 2+H, and 3+H (written using additive notation since this is an additive group). Together they partition the entire group G into equal-size, non-overlapping sets. The index [G : H] is 4.

Cosets and Lagrange's Theorem

Given a subgroup *H* and some *a* in G, we define the left coset *aH* = {*ah* : *h* in *H*}. Because *a* is invertible, the map $\varphi : H \to aH$ given by $\varphi(h) = ah$ is a bijection. Furthermore, every element of *G* is contained in precisely one left coset of H; the left cosets are the equivalence classes corresponding to the equivalence relation $a_1 \sim a_2$ if and only if $a_1^{-1}a_2$ is in H. The number of left cosets of H is called the index of H in G and is denoted by $[G:H]$.

Lagrange's theorem states that for a finite group G and a subgroup H,

$$[G:H] = \frac{|G|}{|H|}$$

where $|G|$ and $|H|$ denote the orders of G and H, respectively. In particular, the order of every subgroup of G (and the order of every element of G) must be a divisor of $|G|$.

Right cosets are defined analogously: $Ha = \{ha : h$ in $H\}$. They are also the equivalence classes for a suitable equivalence relation and their number is equal to $[G:H]$.

If $aH = Ha$ for every a in G, then H is said to be a normal subgroup. Every subgroup of index 2 is normal: the left cosets, and also the right cosets, are simply the subgroup and its complement. More generally, if p is the lowest prime dividing the order of a finite group G, then any subgroup of index p (if such exists) is normal.

To proceed further, we study the notion of subgroup of a given group. That is, if $(G, *)$ is a group and H is a non-empty subset of G then under what condition is $(H, *)$ a group in its own right (it is important to note that the binary operation is the same as that in G). Formally, we have the following definition.

Definition: (Subgroup). Let $(G, *)$ be a group. Then a non-empty subset H of G is said to be a subgroup of G, if H itself forms a group with respect to the binary operation $*$.

Example: 1. Let G be a group with identity element e. Then G and {e} are themselves groups and hence they are subgroups of G. These two subgroups are called trivial subgroups.

2. $\mathbb{Z}$, the set of integers, and $\mathbb{Q}$, the set of rational numbers, are subgroups of $(\mathbb{R}, +)$, the set of real numbers with respect to addition.

3. The set $\{e, r^2, f, r^2f\}$ forms a subgroup of D_4.

4. Let $\sigma \in S_4$. Then, using Theorem, we know that σ has a cycle representation. With this understanding it can be easily verified that the group D_4 is a subgroup of S_4.

5. Consider $H = \{e, r, r^2, \ldots, r^{n-1}\}$ as a subset of D_n. Then it can be easily verified that H is a subgroup of D_4. This subgroup is also written as <r> to indicate that it is generated by the element r of D_4.

Before proceeding further, let us look at the following two results which help us in proving "whether or not a given non-empty subset H of a group G is a subgroup of G"?

Theorem (Subgroup Test) Let G be a group and let H be a non-empty subset of G. Then H is a subgroup of G if for each a, b $\in$ H, $ab^{-1} \in$ H.

Proof. As H is non-empty, we can find an $x \in H$. Therefore, for $a = x$ and $b = x$, the condition $ab^{-1} \in$ H implies that e = $aa^{-1} \in$ H. Thus, H has the identity element of G. Hence, for each h $\in$ H $\subset$ G, eh = h = he.

We now need to prove that for each h ∈ H, h⁻¹ ∈ H. To do so, note that for a = e and b = h the condition ab⁻¹ ∈ H reduces to h⁻¹ = eh⁻¹ ∈ H.

As a third step, we show that H is closed with respect to the binary operation of G. So, let us assume that $x, y \in H$. Then by the previous paragraph, y⁻¹ ∈ H. Therefore, for $a = x$ and $b = y^{-1}$ the condition ab⁻¹ ∈ H implies that $xy = x\left(y^{-1}\right)^{-1} \in H$. Hence, H is also closed with respect to the binary operation of G.

Finally, we see that since the binary operation of H is same as that of G and since associativity holds in G, it holds in H as well.

We now give another result without proof that helps us in deciding whether a non-empty subset of a group is a subgroup or not.

Theorem [Two-Step Subgroup Test] Let H be a non-empty subset of a group G. Then H is a subgroup if the two conditions given below hold.

1. For each a, b ∈ H, ab ∈ H (i.e., H is closed with respect to the binary operation of G).

2. For each a ∈ H, a⁻¹ ∈ H.

We now give a few examples to understand the above theorems.

Example: 1. Consider the group $(\mathbb{Z},+)$. Then in the following cases, the given subsets do not form a subgroup.

(a) Let $H = \{0, 1, 2, 3, \ldots\} \subset \mathbb{Z}$. Note that, for each a, b ∈ H, a + b ∈ H and the identity element 0 ∈ H. But H is not a subgroup of $\mathbb{Z}$, as for all $n \neq 0, -n \notin H$.

(b) Let $H = \mathbb{Z} \setminus \{0\} = \{\ldots, -3, -2, -1, 1, 2, 3, \ldots\} \subset \mathbb{Z}$. Note that, the identity element $0 \notin H$ and hence H is not a subgroup of $\mathbb{Z}$.

(c) Let $H = \{-1, 0, 1\} \subset \mathbb{Z}$. Then H contains the identity element 0 of $\mathbb{Z}$ and for each h ∈ H, h⁻¹ = –h ∈ H. But H is not a subgroup of $\mathbb{Z}$ as $1 + 1 = 2 \notin H$.

2. Let G be an abelian group with identity e. Consider the sets $H = \{x \in G : x^2 = e\}$ and $K = \{x^2 : x \in G\}$. Then prove that both H and K are subgroups of G.

Proof. Clearly e ∈ H and e ∈ K. Hence, both H and K are non-empty subsets of G. We first show that H is a subgroup of G.

As H is non-empty, pick $x, y \in H$. Thus, $x^2 = e = y^2$. We will now use Theorem (Subgroup Test), to show that $xy^{-1} \in H$. But this is equivalent to showing that $\left(xy^{-1}\right)^2 = e$. But this is clearly true as G is abelian implies that

$$\left(xy^{-1}\right)^2 = x^2\left(y^{-1}\right)^2 = e\left(y^2\right)^{-1} = e^{-1} = e.$$

Thus, H is indeed a subgroup of G by Theorem (Subgroup Test).

Now, let us prove that K is a subgroup of G. We have already seen that K is non-empty. Thus, we just need to show that for each $x, y \in K$, $xy^{-1} \in K$.

Note that $x, y \in K$ implies that there exists a, b $\in$ G such that $x = a^2$ and $y = b^2$. As b $\in$ G, b^{-1} $\in$ G. Also, $xy^{-1} = a^2 \left(b^2\right)^{-1} = a^2 \left(b^{-1}\right)^2 = \left(ab^{-1}\right)^2 \in K$ as G is abelian and ab^{-1} $\in$ G. Thus, K is also a subgroup of G.

As a last result of this section, we prove that the condition of the above theorems can be weakened if we assume that H is a finite, non-empty subset of a group G.

Theorem [Finite Subgroup Test] Let G be a group and let H be a non-empty finite subset of G. If H is closed with respect to the binary operation of G then H is a subgroup of G.

Proof. By Theorem [Two-Step Subgroup Test], we just need to show that for each a $\in$ H, a^{-1} $\in$ H. If a = e $\in$ H then a^{-1} = e^{-1} = e $\in$ H. So, let us assume that $a \neq e$ and a $\in$ H. Now consider the set S = {a, a^2, a^3, . . . , a^n, . . .}. As H is closed with respect to the binary operation of G, S $\subset$ H. But H has only finite number of elements. Hence, all these elements of S cannot be distinct. That is, there exist positive integers, say m, n with m > n, such that a^m = a^n. Thus, using Remark (1), one has a^{m-n} = e. Hence, a^{-1} = a^{m-n-1} and by definition a^{m-n-1} $\in$ H.

Lagrange's Theorem (Group Theory)

Lagrange's theorem, in the mathematics of group theory, states that for any finite group G, the order (number of elements) of every subgroup H of G divides the order of G. The theorem is named after Joseph-Louis Lagrange.

Proof of Lagrange's Theorem

This can be shown using the concept of left cosets of H in G. The left cosets are the equivalence classes of a certain equivalence relation on G and therefore form a partition of G. Specifically, x and y in G are related if and only if there exists h in H such that $x = yh$. If we can show that all cosets of H have the same number of elements, then each coset of H has precisely $|H|$ elements. We are then done since the order of H times the number of cosets is equal to the number of elements in G, thereby proving that the order of H divides the order of G. Now, if aH and bH are two left cosets of H, we can define a map $f: aH \rightarrow bH$ by setting $f(x) = ba^{-1}x$.

This map is bijective because its inverse is given by $f^{-1}(y) = ab^{-1}y$.

This proof also shows that the quotient of the orders $|G| \, / \, |H|$ is equal to the index $[G : H]$ (the number of left cosets of H in G). If we allow G and H to be infinite, and write this statement as

$$|G| = [G:H] \cdot |H|,$$

then, seen as a statement about cardinal numbers, it is equivalent to the axiom of choice.

Applications

A consequence of the theorem is that the order of any element a of a finite group (i.e. the smallest positive integer number k with $a^k = e$, where e is the identity element of the group) divides the order of that group, since the order of a is equal to the order of the cyclic subgroup generated by a. If the group has n elements, it follows

$$a^n = e.$$

This can be used to prove Fermat's little theorem and its generalization, Euler's theorem. These special cases were known long before the general theorem was proved.

The theorem also shows that any group of prime order is cyclic and simple. This in turn can be used to prove Wilson's theorem, that if p is prime then p is a factor of $(p-1)!+1$.

Lagrange's theorem can also be used to show that there are infinitely many primes: if there was a largest prime p, then a prime divisor q of the Mersenne number $2^p -1$ would be such that the order of 2 in the multiplicative group $((\mathbb{Z}/q)\setminus 0, \cdot)$ would divide the order of this group which is $q-1$. Hence $p < q$, contradicting the assumption that p is the largest prime.

Existence of Subgroups of Given Order

Lagrange's theorem raises the converse question as to whether every divisor of the order of a group is the order of some subgroup. This does not hold in general: given a finite group G and a divisor d of $|G|$, there does not necessarily exist a subgroup of G with order d. The smallest example is the alternating group $G = A_4$, which has 12 elements but no subgroup of order 6. A *CLT group* is a finite group with the property that for every divisor of the order of the group, there is a subgroup of that order. It is known that a CLT group must be solvable and that every supersolvable group is a CLT group: however there exist solvable groups that are not CLT (for example A_4, the alternating group of degree 4) and CLT groups that are not supersolvable (for example S_4, the symmetric group of degree 4).

There are partial converses to Lagrange's theorem. For general groups, Cauchy's theorem guarantees the existence of an element, and hence of a cyclic subgroup, of order any prime dividing the group order; Sylow's theorem extends this to the existence of a subgroup of order equal to the maximal power of any prime dividing the group order. For solvable groups, Hall's theorems assert the existence of a subgroup of order equal to any unitary divisor of the group order (that is, a divisor coprime to its cofactor).

History

Lagrange did not prove Lagrange's theorem in its general form. He stated, in his article *Réflexions sur la résolution algébrique des équations*, that if a polynomial in n variables has its variables permuted in all $n!$ ways, the number of different polynomials that are obtained is always a factor of $n!$. (For example, if the variables x, y, and z are permuted in all 6 possible ways in the polynomial $x + y - z$ then we get a total of 3 different polynomials: $x + y - z$, $x + z - y$, and $y + z - x$. Note that 3 is a factor of 6.) The number of such polynomials is the index in the symmetric group S_n of the subgroup H of permutations that preserve the polynomial. (For the example of $x + y - z$, the subgroup

H in S_3 contains the identity and the transposition (xy).) So the size of H divides $n!$. With the later development of abstract groups, this result of Lagrange on polynomials was recognized to extend to the general theorem about finite groups which now bears his name.

In his *Disquisitiones Arithmeticae* in 1801, Carl Friedrich Gauss proved Lagrange's theorem for the special case of $Z(p)^*$, the multiplicative group of nonzero integers modulo p, where p is a prime. In 1844, Augustin-Louis Cauchy proved Lagrange's theorem for the symmetric group S_n.

Camille Jordan finally proved Lagrange's theorem for the case of any permutation group in 1861.

In this section, we prove the first fundamental theorem for groups that have finite number of elements. To do so, we start with the following example to motivate our definition and the ideas that they lead to.

Example: Consider the set $\mathbb{R}^2 = \{(x, y): x, y \in \mathbb{R}\}$. Then $\mathbb{R}^2$ is an abelian group with respect to component wise addition. That is, for each $(x_1, y_1), (x_2, y_2) \in \mathbb{R}^2$, the binary operation is defined by $(x_1, y_1) + (x_2, y_2) = (x_1 + x_2, y_1 + y_2)$. Check that if H is a non-trivial subgroup of $\mathbb{R}^2$ then H represents a line passing through (0, 0).

Hence, $H_1 = \{(x, y) \in \mathbb{R}^2 : y = 0\}, H_2 = \{(x, y) \in \mathbb{R}^2 : x = 0\}$ and $H_3 = \{(x, y) \in \mathbb{R}^2 : y = 3x\}$ are subgroups of $\mathbb{R}^2$. Note that H_1 represents the X-axis, H_2 represents the Y-axis and H_3 represents a line passes through the origin and has slope 3.

Fix the element $(2, 3) \in \mathbb{R}^2$. Then

1 $(2, 3) + H_1 = \{(2,3) + (x, y) : y = 0\} = \{(2 + x, 3) : x \in \mathbb{R}\}$. This is the equation of a line that passes through the point (2, 3) and is parallel to the X-axis.

2. verify that $(2, 3) + H_2$ represents a line that passes through the point (2, 3) and is parallel to the Y-axis.

3. $(2, 3) + H_3 = \{(2 + x, 3 + 3x) : x \in \mathbb{R}\} = \{(x, y) \in \mathbb{R}^2 : y = 3x - 3\}$. So, this represents a line that has slope 3 and passes through the point (2, 3).

So, we see that if we fix a subgroup H of $\mathbb{R}^2$ and take any point $(x_0, y_0) \in \mathbb{R}^2$, then the set $(x_0, y_0) + H$ gives a line that is a parallel shift of the line represented by H and $(x_0, y_0) + H$ contains the point (x_0, y_0). Hence, it can be easily observed that

1. (x_1, y_1) lies on the line $(x_0, y_0) + H$ if and only if $(x_0, y_0) + H = (x_1, y_1) + H$.

2. for any two $(x_0, y_0), (x_1, y_1) \in \mathbb{R}^2$, either $(x_0, y_0) + H = (x_1, y_1) + H$ or they represent two parallel lines which themselves are parallel to the line represented by H.

3. $\bigcup_{x \in \mathbb{R}} \bigcup_{y \in \mathbb{R}} (x, y) + H = \mathbb{R}^2$

That is, if we define a relation, denoted $\sim$, in $\mathbb{R}^2$ by $(x_1, y_1) \sim (x_2, y_2)$, whenever $(x_1 - x_2, y_1 - y_2) \in H$, then the above observations imply that this relation is an equivalence relation. Hence, as (x, y) vary over all the points of $\mathbb{R}^2$, we get a partition of $\mathbb{R}^2$. Moreover, the equivalence class containing the point (x_0, y_0) is the set $(x_0, y_0) + H$.

Therefore, we see that given a subgroup H of a group G, it may be possible to partition the group G into subsets that are in some sense similar to H itself. Example 3.4.1 also implies that for each g ∈ G, we need to consider the set g + H, if G is an additive group or either the set gH or the set Hg, if G is a multiplicative group. So, we are led to the following definition.

Definition (Left and Right Coset) Let G be a group and let H be a subgroup of G. Then for each g ∈ G the set

1. gH = {gh : h ∈ H} is called the left coset of H in G.

2. Hg = {hg : h ∈ H} is called the right coset of H in G.

Remark: Since the identity element e ∈ H, for each fixed g ∈ G, g = ge ∈ gH. Hence, we often say that gH is the left coset of H containing g. Similarly, g ∈ Hg and hence Hg is said to be the right coset of H containing g.

Example: Consider the group D_4 and let H = {e, f } and K = {e, r²} be two subgroups of D_4. Then observe the following:

$$H = \{e, f\} = Hf, Hr = \{r, fr\} = H\ fr,$$
$$H\ r^2 = \{r^2, fr^2\} = H\ fr^2 \quad and \quad H\ r^3 = \{r^3, fr^3\} = H\ fr^3. \tag{1}$$

$$H = \{e, f\} = fH, rH = \{r, rf\} = rf\ H,$$
$$r^2H = \{r^2, r^2 f\} = r^2 f\ H \quad and \quad r^3H = \{r^3, r^3 f\} = r^3 f\ H. \tag{2}$$

$$K\ = \{e, r^2\} = K\ r^2 = r^2K, Kr\ = \{r, r^3\} = rK = Kr^3 = r^3K$$
$$Kf = \{f, r^2 f\ \} = f\ K = K\ r^2 f = r^2 f\ K \quad and \tag{3}$$
$$K\ fr = \{fr, fr^3\} = fr\ K = K\ fr^3 = fr^3K.$$

From (1) and (2), we note that in general $Hg \neq gH$, for each g ∈ D_4, whereas from (3), we see that Kg = gK, for each g ∈ D_4. So, there should be a way to distinguish between these two subgroups of D_4. This leads to study of normal subgroups and beyond. The interested reader can look at any standard book in abstract algebra to go further in this direction.

Now, let us come back to the partition of a group using its subgroup. The proof of the theorem is left as an exercise for the readers.

Theorem: Let H be a subgroup of a group G. Suppose a, b ∈ G. Then the following results hold for left cosets of H in G:

1. aH = H if and only if a ∈ H,

2. aH is a subgroup of G if and only if a ∈ H,

3. either $aH = bH\ or\ aH \cap bH = \emptyset$,

4. aH = bH if and only if a⁻¹b ∈ H.

Similarly one obtains the following results for right cosets of H in G.

1. Ha = H if and only if a ∈ H,

2. Ha is a subgroup of G if and only if a ∈ H,

3. either $Ha = Hb \ or \ Ha \cap Hb = \emptyset$,

4. Ha = Hb if and only if ab^{-1} ∈ H.

Furthermore, aH = Ha if and only if H = aHa^{-1} = {aha^{-1} : h ∈ H}.

To proceed further, we need the following definition.

Definition (Order of a Group) The number of elements in G, denoted |G|, is called the order of G. If |G| < ∞, then G is called a group of finite order.

We are now ready to prove the main result of this section, namely the Lagrange's Theorem.

Theorem: Let H be a subgroup of a finite group G. Then |H| divides |G|. Moreover, the number of distinct left (right) cosets of H in G equals $\dfrac{|G|}{|H|}$.

Proof. We give the proof for left cosets. A similar proof holds for right cosets. Since G is a finite group, the number of left cosets of H in G is finite. Let $g_1H, g_2H, \ldots, g_mH$ be the collection of all left cosets of H in G. Then by Theorem, two cosets are either equal or they are disjoint. That is, G is a disjoint union of the sets $g_1H, g_2H, \ldots, g_mH$.

Also, it can be easily verified that |aH| = |bH|, for each a, b ∈ G. Hence, $|g_iH| = |H|$, for all i = 1, 2, . . ., m. Thus, $|G| = \left| \bigcup\limits_{i=1}^{m} g_iH \right| = \sum\limits_{i=1}^{m} |g_iH| = m|H|$ (the disjoint union gives the second equality). Thus,

|H| divides |G| and the number of left cosets equals $m = \dfrac{|G|}{|H|}$.

Remark: The number m in the above Theorem is called the index of H in G, and is denoted by [G : H] or $i_G(H)$.

Hence, Theorem is a statement about any subgroup of a finite group. It may so happen that the group G and its subgroup H may have infinite number of elements but the number of left (right) cosets of H in G may be finite. One still talks of index of H in G in such cases. For example, consider $H = 10\mathbb{Z}$ as a subgroup of the additive group $\mathbb{Z}$. Then $[\mathbb{Z}: H] = 10$. In general, for a fixed positive integer m, consider the subgroup $m\mathbb{Z}$ of the additive group $\mathbb{Z}$. Then it can be easily verified that $[\mathbb{Z}: m\mathbb{Z}] = m$.

Applications of Lagrange's Theorem

Definition (Order of an Element) Let G be a group and let g ∈ G. Then the smallest positive integer m such that g^m = e is called the order of g. If there is no such positive integer then g is said to have infinite order. The order of an element is denoted by O(g).

Example: 1. The only element of order 1 in a group G is the identity element of G.

2. In D_4, the elements r^2, f, r f, r^2f, r^3f have order 2, whereas the elements r and r^3 have order 4.

With the definition of the order of an element, we now prove that in general, the converse of Lagrange's Theorem is not true. To see this consider the group G discussed in Example 3.2.1.2a. This group has 12 elements and 6 divides 12. Whereas it can be shown that G doesn't have a subgroup of order 6. We give a proof for better understanding of cosets.

Proof. Let if possible, H be a subgroup of order 6 in G, where

G = {e, (234), (243), (124), (142), (123), (132), (134), (143), (12)(34), (13)(24), (14)(23)}.

Observe that G has exactly 8 elements of the form (ijk), for distinct numbers i, j and k, and each has order 3. Hence, G has exactly 8 elements of order 3. Let a ∈ G with O(a) = 3. Then using Theorem, we see that cosets of H in G will be exactly 2 and at the same time, the possible cosets could be H, aH and a^2H (as a^3 = e, no other coset exists). Hence, at most two of the cosets H, aH and a^2H are distinct. But, using Theorem, it can be easily verified that the equality of any two of them gives a ∈ H. Therefore, all the 8 elements of order 3 must be elements of H. That is, H must have at least 9 elements (8 elements of order 3 and one identity). This is absurd as |H| = 6.

We now derive some important corollaries of Lagrange's Theorem. We omit the proof as it can be found in any standard textbook in abstract algebra. The first corollary is about the order of an element of a finite group. The observation that for each g ∈ G, the set H = {e, g, g^2, g^3, . . .} forms a subgroup of any finite group G gives the proof of the next result.

Corollary: Let G be a finite group and let g ∈ G. Then O(g) divides |G|.

Remark: Corollary above implies that if G is a finite group of order n then the possible orders of its elements are the divisors of n. For example, if |G| = 30 then for each g ∈ G, O(g) ∈ {1, 2, 3, 5, 6, 10, 15, 30}.

Let G be a finite group. Then in the first corollary, we have shown that for any g ∈ G, O(g) divides |G|. Therefore, |G| = m • O(g), for some positive integer m. Hence

$$g^{|G|} = g^{m \cdot O(g)} = \left(g^{O(g)}\right)^m = e^m = e.$$

This observation gives our next result.

Corollary : Let G be a finite group. Then, for each g ∈ G, $g^{|G|}$ = e.

Let P be an odd prime and consider the set $\mathbb{Z}_p^* = \{1, 2, \ldots, p-1\}$. Then, check that $\mathbb{Z}_p^*$ forms a group with respect to the binary operation

a ⊙ b = the remainder, when ab is divided by p.

Applying above Corollary to $\mathbb{Z}_p^*$ gives the famous result called the Fermat's Little Theorem. To state this, recall that for a, $b \in \mathbb{Z}$, the notation "a ≡ b (mod p)" indicates that p divides a − b.

Corollary: Let a be any positive integer and let p be a prime. Then a^{p-1} ≡ 1 (mod p), if p does not divide a. In general, $a^p \equiv a \pmod{p}$.

We now state without proof a generalization of the Fermat's Little Theorem, popularly known as the Euler's Theorem. To do so, let $U_n = \{k : 1 \le k \le n, \gcd(k, n) = 1\}$, for each positive integer n. Then U_n, with binary operation

$a \odot b$ = the remainder, when ab is divided by n

forms a group. Also, recall that the symbol $\varphi(n)$ gives the number of integers between 1 and n that are coprime to n. That is, $|U_n| = \varphi(n)$, for each positive integer n. Now applying Corollary (2) to U_n, gives the next result.

Corollary: Let a, $n \in \mathbb{Z}$ with n > 0. If gcd(a, n) = 1 then $a^{\varphi(n)} \equiv 1 \pmod{n}$.

Example: 1. Find the unit place in the expansion of 13^{1001}.

Solution: Observe that $13 \equiv 3 \pmod{10}$. So, $13^{1001} \equiv 3^{1001} \pmod{10}$. Also, $3 \in U_{10}$ and therefore by Corollary (2), $3^{|U_{10}|} = 1 \pmod{10}$. But $|U_{10}| = 4$ and $1001 = 4 \cdot 250 + 1$. Thus,

$13^{1001} \equiv 3^{1001} \equiv 3^{4 \cdot 250 + 1} \equiv (3^4)^{250} \cdot 3^1 \equiv 1 \cdot 3 \equiv 3 \pmod{10}$.

Hence, the unit place in the expansion of 13^{1001} is 3.

2. Find the unit and tens place in the expansion of 23^{1002}.

Solution: Observe that $23 \in U_{100}$ and $23^{|U_{100}|} = 1 \pmod{100}$. But $|U_{100}| = 40$ and

$1002 = 40 \cdot 25 + 2$. Hence

$23^{1002} \equiv 23^{40 \cdot 25 + 2} \equiv (23^{40})^{25} \cdot 23^2 \equiv 1 \cdot 23^2 \equiv 529 \equiv 29 \pmod{100}$.

Hence, the unit place is 9 and the tens place is 2 in the expansion of 23^{1002}.

Group Action

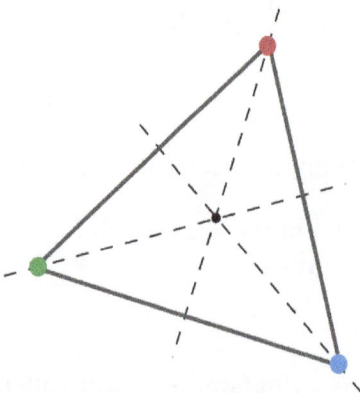

Given an equilateral triangle, the counterclockwise rotation by 120° around the center of the triangle maps every vertex of the triangle to another one. The cyclic group C_3 consisting of the rotations by 0°, 120° and 240° acts on the set of the three vertices.

In mathematics, an action of a group is a way of interpreting the elements of the group as "acting" on some space in a way that preserves the structure of that space. Common examples of spaces that groups act on are sets, vector spaces, and topological spaces. Actions of groups on vector spaces are called representations of the group.

Some groups can be interpreted as acting on spaces in a canonical way. For example, the symmetric group of a finite set consists of all bijective transformations of that set; thus, applying any element of the permutation group to an element of the set will produce another element of the set. More generally, symmetry groups such as the homeomorphism group of a topological space or the general linear group of a vector space, as well as their subgroups, also admit canonical actions. For other groups, an interpretation of the group in terms of an action may have to be specified, either because the group does not act canonically on any space or because the canonical action is not the action of interest. For example, we can specify an action of the two-element cyclic group $C_2 = \{0,1\}$ on the finite set $\{a,b,c\}$ by specifying that 0 (the identity element) sends $a \mapsto a, b \mapsto b, c \mapsto c$, and that 1 sends $a \mapsto b, b \mapsto a, c \mapsto c$. This action is not canonical.

A common way of specifying non-canonical actions is to describe a homomorphism φ from a group G to the group of symmetries of a set X. The action of an element $g \in G$ on a point $x \in X$ is assumed to be identical to the action of its image $\varphi(g) \in \mathrm{Sym}(X)$ on the point x. The homomorphism φ is also frequently called the "action" of G, since specifying φ is equivalent to specifying an action. Thus, if G is a group and X is a set, then an action of G on X may be formally defined as a group homomorphism φ from G to the symmetric group of X. The action assigns a permutation of X to each element of the group in such a way that:

- the identity element of G is assigned the identity transformation of X;

- any product gk of two elements of G is assigned the composition of the permutations assigned to g and k.

If X has additional structure, then φ is only called an action if for each $g \in G$, the permutation $\varphi(g)$ preserves the structure of X.

The abstraction provided by group actions is a powerful one, because it allows geometrical ideas to be applied to more abstract objects. Many objects in mathematics have natural group actions defined on them. In particular, groups can act on other groups, or even on themselves. Because of this generality, the theory of group actions contains wide-reaching theorems, such as the orbit stabilizer theorem, which can be used to prove deep results in several fields.

Definition

If G is a group and X is a set, then a (*left*) *group action* φ of G on X is a function

$$\varphi : G \times X \to X : (g,x) \mapsto \varphi(g,x)$$

that satisfies the following two axioms (where we denote $\varphi(g, x)$ as $g.x$):

Identity

$e.x = x$ for all x in X. (Here, e denotes the identity element of the group G.)

Compatibility

$(gh).x = g.(h.x)$ for all g, h in G and all x in X. (Here, gh denotes the result of applying the group operation of G to the elements g and h.)

The group G is said to act on X (on the left). The set X is called a (*left*) *G-set*.

From these two axioms, it follows that for every g in G, the function which maps x in X to $g.x$ is a bijective map from X to X (its inverse being the function which maps x to $g^{-1}.x$). Therefore, one may alternatively define a group action of G on X as a group homomorphism from G into the symmetric group $\text{Sym}(X)$ of all bijections from X to X.

In complete analogy, one can define a *right group action* of G on X as an operation $X \times G \to X$ mapping (x, g) to $x.g$ and satisfying the two axioms:

Identity

$x.e = x$ for all x in X.

Compatibility

$x.(gh) = (x.g).h$ for all g, h in G and all x in X;

The difference between left and right actions is in the order in which a product like gh acts on x. For a left action h acts first and is followed by g, while for a right action g acts first and is followed by h. Because of the formula $(gh)^{-1} = h^{-1}g^{-1}$, one can construct a left action from a right action by composing with the inverse operation of the group. Also, a right action of a group G on X is the same thing as a left action of its opposite group G^{op} on X. It is thus sufficient to only consider left actions without any loss of generality.

Types of Actions

The action of G on X is called

- *Transitive* if X is non-empty and if for each pair x, y in X there exists a g in G such that $g.x = y$. For example, the action of the symmetric group of X is transitive, the action of the general linear group or the special linear group of a vector space V on $V \setminus \{0\}$ is transitive, but the action of the orthogonal group of a Euclidean space E is not transitive on $E \setminus \{0\}$ (it is transitive on the unit sphere of E, though).

- *Faithful* (or *effective*) if for every two distinct g, h in G there exists an x in X such that $g.x \neq h.x$; or equivalently, if for each $g \neq e$ in G there exists an x in X such that $g.x \neq x$. In other words, in a faithful group action, different elements of G induce different permutations of X. In algebraic terms, a group G acts faithfully on X if and only if the corresponding homomorphism to the symmetric group, $G \to \text{Sym}(X)$, has a trivial kernel. Thus, for a faithful action, G embeds into a permutation group on X; specifically, G is isomorphic to its image in $\text{Sym}(X)$. If G does not act faithfully on X, one can easily modify the group to obtain a faithful action. If we define $N = \{g$ in $G : g.x = x$ for all x in $X\}$, then N is a normal subgroup of G; indeed, it is the kernel of the homomorphism $G \to \text{Sym}(X)$. The factor group G/N acts faithfully on X by setting $(gN).x = g.x$. The original action of G on X is faithful if and only if $N = \{e\}$.

- *Free* (or *semiregular* or *fixed point free*) if, given g, h in G, the existence of an x in X with $g.x = h.x$ implies $g = h$. Equivalently: if g is a group element and there exists an x in X with $g.x = x$ (that is, if g has at least one fixed point), then g is the identity. Note that a free action on a non-empty set is faithful.

- *Regular* (or *simply transitive* or *sharply transitive*) if it is both transitive and free; this is equivalent to saying that for every two x, y in X there exists precisely one g in G such that $g.x = y$. In this case, X is called a principal homogeneous space for G or a G-torsor. The action of any group G on itself by left multiplication is regular, and thus faithful as well. Every group can, therefore, be embedded in the symmetric group on its own elements, Sym(G). This result is known as Cayley's theorem.

- *n-transitive* if X has at least n elements and for all pairwise distinct $x_1, ..., x_n$ and pairwise distinct $y_1, ..., y_n$ there is a g in G such that $g.x_k = y_k$ for $1 \leq k \leq n$. A 2-transitive action is also called *doubly transitive*, a 3-transitive action is also called *triply transitive*, and so on. Such actions define interesting classes of subgroups in the symmetric groups: 2-transitive groups and more generally multiply transitive groups. The action of the symmetric group on a set with n elements is always n-transitive; the action of the alternating group is n-2-transitive.

- *Sharply n-transitive* if there is exactly one such g.

- *Primitive* if it is transitive and preserves no non-trivial partition of X.

- *Locally free* if G is a topological group, and there is a neighbourhood U of e in G such that the restriction of the action to U is free; that is, if $g.x = x$ for some x and some g in U then $g = e$.

Furthermore, if G acts on a topological space X, then the action is:

- *Wandering* if every point $x \in X$ has a neighbourhood U such that $\{g \in G : g \cdot U \cap U \neq \varnothing\}$ is finite. For example, the action of $\mathbb{Z}^n$ on $\mathbb{R}^n$ by translations is wandering. The action of the modular group on the Poincaré half-plane is also wandering.

- *Properly discontinuous* if X is a locally compact space and for every compact subset $K \subset X$ the set $\{g \in G : gK \cap K \neq \varnothing\}$ is finite. The wandering actions given above are also properly discontinuous. On the other hand, the action of $\mathbb{Z}$ on $\mathbb{R}^2 \setminus \{0\}$ by the linear map $(x, y) \mapsto (2x, 1/2y)$ is wandering and free but not properly discontinuous.

- *Proper* if G is a topological group and the map from $G \times X \rightarrow X \times X : (g, x) \mapsto (g \cdot x, x)$ is proper. If G is discrete then properness is equivalent to proper discontinuity for G-actions.

- Said to have *discrete orbits* if the orbit of each $x \in X$ under the action of G is discrete in X.

If X is a non-zero module over a ring R and the action of G is R-linear then it is said to be

- *Irreducible* if there is no nonzero proper invariant submodule.

Orbits and Stabilizers

Consider a group G acting on a set X. The *orbit* of an element x in X is the set of elements in X to which x can be moved by the elements of G. The orbit of x is denoted by $G.x$:

In the compound of five tetrahedra, the symmetry group is the (rotational) icosahedral group I of order 60, while the stabilizer of a single chosen tetrahedron is the (rotational) tetrahedral group T of order 12, and the orbit space I/T (of order $60/12 = 5$) is naturally identified with the 5 tetrahedra – the coset gT corresponds to the tetrahedron to which g sends the chosen tetrahedron.

$$G.x = \{ g.x \mid g \in G \}.$$

The defining properties of a group guarantee that the set of orbits of (points x in) X under the action of G form a partition of X. The associated equivalence relation is defined by saying $x \sim y$ if and only if there exists a g in G with $g.x = y$. The orbits are then the equivalence classes under this relation; two elements x and y are equivalent if and only if their orbits are the same; i.e., $G.x = G.y$.

The group action is transitive if and only if it has only one orbit, i.e., if there exists x in X with $G.x = X$. This is the case if and only if $G.x = X$ for *all* x in X.

The set of all orbits of X under the action of G is written as X/G (or, less frequently: $G\backslash X$), and is called the *quotient* of the action. In geometric situations it may be called the *orbit space*, while in algebraic situations it may be called the space of *coinvariants*, and written X_G, by contrast with the invariants (fixed points), denoted X^G: the coinvariants are a *quotient* while the invariants are a *subset*. The coinvariant terminology and notation are used particularly in group cohomology and group homology, which use the same superscript/subscript convention.

Invariant Subsets

If Y is a subset of X, we write GY for the set $\{ g.y : y \in Y$ and $g \in G \}$. We call the subset Y *invariant under G* if $G.Y = Y$ (which is equivalent to $G.Y \subseteq Y$). In that case, G also operates on Y by restricting the action to Y. The subset Y is called *fixed under G* if $g.y = y$ for all g in G and all y in Y. Every subset that is fixed under G is also invariant under G, but not vice versa.

Every orbit is an invariant subset of X on which G acts transitively. The action of G on X is *transitive* if and only if all elements are equivalent, meaning that there is only one orbit.

A *G-invariant* element of X is $x \in X$ such that $g.x = x$ for all $g \in G$. The set of all such x is denoted X^G and called the *G-invariants* of X. When X is a G-module, X^G is the zeroth group cohomology group of G with coefficients in X, and the higher cohomology groups are the derived functors of the functor of G-invariants.

Fixed Points and Stabilizer Subgroups

Given g in G and x in X with $g.x = x$, we say x is a fixed point of g and g fixes x.

For every x in X, we define the *stabilizer subgroup* of G with respect to x (also called the *isotropy group* or *little group*) as the set of all elements in G that fix x:

$$G_x = \{ g \in G \mid g \cdot x = x \}.$$

This is a subgroup of G, though typically not a normal one. The action of G on X is free if and only if all stabilizers are trivial. The kernel N of the homomorphism with the symmetric group, $G \rightarrow \mathrm{Sym}(X)$, is given by the intersection of the stabilizers G_x for all x in X. If N is trivial, the action is said to be faithful (or effective).

Let x and y be two elements in X, and let g be a group element such that $y = g.x$. Then the two stabilizer groups G_x and G_y are related by $G_y = g\, G_x\, g^{-1}$. Proof: by definition, $h \in G_y$ if and only if $h.(g.x) = g.x$. Applying g^{-1} to both sides of this equality yields $(g^{-1}hg).x = x$; that is, $g^{-1}hg \in G_x$.

The above says that the stabilizers of elements in the same orbit are conjugate to each other. Thus, to each orbit, one can associate a conjugacy class of a subgroup of G (i.e., the set of all conjugates of the subgroup). Let (H) denote the conjugacy class of H. Then one says that the orbit O has type (H) if the stabilizer   of some/any x in O belongs to (H). A maximal orbit type is often called a principal orbit type.

Orbit-stabilizer Theorem and Burnside's Lemma

Orbits and stabilizers are closely related. For a fixed x in X, consider the map from G to X given by $g \mapsto g.x$ for all $g \in G$. The image of this map is the orbit of x and the coimage is the set of all left cosets of G_x. The standard quotient theorem of set theory then gives a natural bijection between G/G_x and $G.x$. Specifically, the bijection is given by $hG_x \mapsto h.x$. This result is known as the *orbit-stabilizer theorem*. From a more categorical perspective, the orbit-stabilizer theorem comes from the fact that every G-set is a sum of quotients of the G-set G.

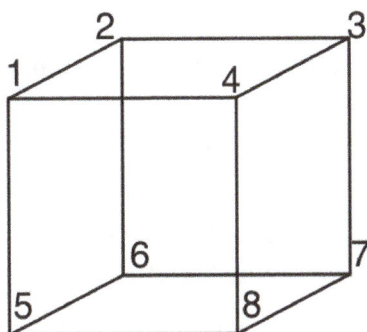

Cubical graph with vertices labeled.

If G is finite then the orbit-stabilizer theorem, together with Lagrange's theorem, gives

$$|G.x| = [G{:}G_x] = |G| / |G_x|.$$

This result is especially useful since it can be employed for counting arguments (typically in situations where X is finite as well).

Example: One can use the orbit-stabilizer theorem to count the automorphisms of a graph. Consider the cubical graph as pictured, and let G denote its automorphism group. Then G acts on the set of vertices $\{1, 2, \ldots, 8\}$, and this action is transitive as can be seen by composing rotations about the center of the cube. Thus, by the orbit-stabilizer theorem, we have that $|G| = |G.1||G_1| = 8|G_1|$. Applying the theorem now to the stabilizer G_1, we obtain $|G_1| = |(G_1).2||(G_1)_2|$. Any element of G that fixes 1 must send 2 to either 2, 4 or 5. There are such automorphisms; consider for example the map that transposes 2 and 4, transposes 6 and 8, and fixes the other vertices. Thus, $|(G_1).2| = 3$. Applying the theorem a third time gives $|(G_1)_2| = |((G_1)_2).3||((G_1)_2)_3|$. Any element of G that fixes 1 and 2 must send 3 to either 3 or 6, and one easily finds such automorphisms. Thus, $|((G_1)_2).3| = 2$. One also sees that $((G_1)_2)_3$ consists only of the identity automorphism, as any element of G fixing 1, 2 and 3 must also fix 4 and consequently all other vertices. Combining the preceding calculations, we now obtain $|G| = 8 \cdot 3 \cdot 2 \cdot 1 = 48$.

A result closely related to the orbit-stabilizer theorem is Burnside's lemma:

$$|X/G| = \frac{1}{|G|} \sum_{g \in G} |X^g|,$$

where X^g is the set of points fixed by g. This result is mainly of use when G and X are finite, when it can be interpreted as follows: the number of orbits is equal to the average number of points fixed per group element.

Fixing a group G, the set of formal differences of finite G-sets forms a ring called the Burnside ring of G, where addition corresponds to disjoint union, and multiplication to Cartesian product.

Examples

- The *trivial* action of any group G on any set X is defined by $g.x = x$ for all g in G and all x in X; that is, every group element induces the identity permutation on X.

- In every group G, left multiplication is an action of G on G: $g.x = gx$ for all g, x in G.

- In every group G with subgroup H, left multiplication is an action of G on the set of cosets G/H: $g.aH = gaH$ for all g,a in G. In particular if H contains no nontrivial normal subgroups of G this induces an isomorphism from G to a subgroup of the permutation group of degree $[G : H]$.

- In every group G, conjugation is an action of G on G: $g.x = gxg^{-1}$. An exponential notation is commonly used for the right-action variant: $x^g = g^{-1}xg$; it satisfies $(x^g)^h = x^{gh}$.

- In every group G with subgroup H, conjugation is an action of G on conjugates of H: $g.K = gKg^{-1}$ for all g in G and K conjugates of H.

- The symmetric group S_n and its subgroups act on the set $\{\,1, ..., n\,\}$ by permuting its elements.

- The symmetry group of a polyhedron acts on the set of vertices of that polyhedron. It also acts on the set of faces or the set of edges of the polyhedron.

- The symmetry group of any geometrical object acts on the set of points of that object.

- The automorphism group of a vector space (or graph, or group, or ring...) acts on the vector space (or set of vertices of the graph, or group, or ring...).

- The general linear group $GL(n, K)$ and its subgroups, particularly its Lie subgroups (including the special linear group $SL(n, K)$, orthogonal group $O(n, K)$, special orthogonal group $SO(n, K)$, and symplectic group $Sp(n, K)$) are Lie groups that act on the vector space K^n. The group operations are given by multiplying the matrices from the groups with the vectors from K^n.

- The affine group acts transitively on the points of an affine space, and the subgroup V of the affine group (i.e., a vector space) transitive and free (i.e., *regular*) action on these points; indeed this can be used to give a definition of an affine space.

- The projective linear group $PGL(n+1, K)$ and its subgroups, particularly its Lie subgroups, which are Lie groups that act on the projective space $P^n(K)$. This is a quotient of the action of the general linear group on projective space. Particularly notable is $PGL(2, K)$, the symmetries of the projective line, which is sharply 3-transitive, preserving the cross ratio; the Möbius group $PGL(2, C)$ is of particular interest.

- The isometries of the plane act on the set of 2D images and patterns, such as wallpaper patterns. The definition can be made more precise by specifying what is meant by image or pattern; e.g., a function of position with values in a set of colors. Isometries are in fact one example of affine group (action).

- The sets acted on by a group G comprise the category of G-sets in which the objects are G-sets and the morphisms are G-set homomorphisms: functions $f : X \to Y$ such that $g.(f(x)) = f(g.x)$ for every g in G.

- The Galois group of a field extension L/K acts on the field L but has only a trivial action on elements of the subfield K. Subgroups of $Gal(L/K)$ correspond to subfields of L that contain K, i.e., intermediate field extensions between L and K.

- The additive group of the real numbers (R, +) acts on the phase space of "well-behaved" systems in classical mechanics (and in more general dynamical systems) by time translation: if t is in R and x is in the phase space, then x describes a state of the system, and $t + x$ is defined to be the state of the system t seconds later if t is positive or $-t$ seconds ago if t is negative.

- The additive group of the real numbers (R, +) acts on the set of real functions of a real variable in various ways, with $(t.f)(x)$ equal to, e.g., $f(x + t)$, $f(x) + t$, $f(xe^t)$, $f(x)e^t$, $f(x + t)e^t$, or $f(xe^t) + t$, but not $f(xe^t + t)$.

- Given a group action of G on X, we can define an induced action of G on the power set of X, by setting $g.U = \{g.u : u \in U\}$ for every subset U of X and every g in G. This is useful, for instance, in studying the action of the large Mathieu group on a 24-set and in studying symmetry in certain models of finite geometries.

- The quaternions with norm 1 (the versors), as a multiplicative group, act on R³: for any such quaternion $z = \cos \alpha/2 + v \sin \alpha/2$, the mapping $f(x) = zxz^*$ is a counterclockwise rotation through an angle α about an axis given by a unit vector v; z is the same rotation.

Group Actions and Groupoids

The notion of group action can be put in a broader context by using the *action groupoid* $G' = G \ltimes X$ associated to the group action, thus allowing techniques from groupoid theory such as presentations and fibrations. Further the stabilizers of the action are the vertex groups, and the orbits of the action are the components, of the action groupoid.

This action groupoid comes with a morphism $p : G' \rightarrow G$ which is a *covering morphism of groupoids*. This allows a relation between such morphisms and covering maps in topology.

Morphisms and Isomorphisms Between *G*-sets

If X and Y are two G-sets, we define a *morphism* from X to Y to be a function $f : X \rightarrow Y$ such that $f(g.x) = g.f(x)$ for all g in G and all x in X. Morphisms of G-sets are also called *equivariant maps* or *G-maps*.

The composition of two morphisms is again a morphism.

If a morphism f is bijective, then its inverse is also a morphism, and we call f an *isomorphism* and the two G-sets X and Y are called *isomorphic*; for all practical purposes, they are indistinguishable in this case.

Some example isomorphisms:

- Every regular G action is isomorphic to the action of G on G given by left multiplication.

- Every free G action is isomorphic to $G \times S$, where S is some set and G acts on $G \times S$ by left multiplication on the first coordinate. (S can be taken to be the set of orbits X/G.)

- Every transitive G action is isomorphic to left multiplication by G on the set of left cosets of some subgroup H of G. (H can be taken to be the stabilizer group of any element of the original G-set.the original action.)

With this notion of morphism, the collection of all G-sets forms a category; this category is a Grothendieck topos (in fact, assuming a classical metalogic, this topos will even be Boolean).

Continuous Group Actions

One often considers *continuous group actions*: the group G is a topological group, X is a topological space, and the map $G \times X \rightarrow X$ is continuous with respect to the product topology of $G \times X$. The space X is also called a *G-space* in this case. This is indeed a generalization, since every group can be considered a topological group by using the discrete topology. All the concepts introduced above still work in this context, however we define morphisms between G-spaces to be *continuous* maps compatible with the action of G. The quotient X/G inherits the quotient topology from X, and is called the *quotient space* of the action. The above statements about isomorphisms for regular, free and transitive actions are no longer valid for continuous group actions.

If X is a regular covering space of another topological space Y, then the action of the deck transformation group on X is properly discontinuous as well as being free. Every free, properly discontinuous action of a group G on a path-connected topological space X arises in this manner: the quotient map $X \mapsto X/G$ is a regular covering map, and the deck transformation group is the given action of G on X. Furthermore, if X is simply connected, the fundamental group of X/G will be isomorphic to G.

These results have been generalized to obtain the fundamental groupoid of the orbit space of a discontinuous action of a discrete group on a Hausdorff space, as, under reasonable local conditions, the orbit groupoid of the fundamental groupoid of the space. This allows calculations such as the fundamental group of the symmetric square of a space X, namely the orbit space of the product of X with itself under the twist action of the cyclic group of order 2 sending (x, y) to (y, x).

An action of a group G on a locally compact space X is *cocompact* if there exists a compact subset A of X such that $GA = X$. For a properly discontinuous action, cocompactness is equivalent to compactness of the quotient space X/G.

The action of G on X is said to be *proper* if the mapping $G \times X \to X \times X$ that sends $(g, x) \mapsto (g.x, x)$ is a proper map.

Strongly Continuous Group Action and Smooth Points

A group action of a topological group G on a topological space X is said to be *strongly continuous* if for all x in X, the map $g \mapsto g.x$ is continuous with respect to the respective topologies. Such an action induces an action on the space of continuous functions on X by defining $(g,f)(x) = f(g^{-1}.x)$ for every g in G, f a continuous function on X, and x in X. Note that, while every continuous group action is strongly continuous, the converse is not in general true.

The subspace of *smooth points* for the action is the subspace of X of points x such that $g \mapsto g.x$ is smooth; i.e., it is continuous and all derivatives are continuous.

Variants and Generalizations

One can also consider actions of monoids on sets, by using the same two axioms as above. This does not define bijective maps and equivalence relations however.

Instead of actions on sets, one can define actions of groups and monoids on objects of an arbitrary category: start with an object X of some category, and then define an action on X as a monoid homomorphism into the monoid of endomorphisms of X. If X has an underlying set, then all definitions and facts stated above can be carried over. For example, if we take the category of vector spaces, we obtain group representations in this fashion.

One can view a group G as a category with a single object in which every morphism is invertible. A (left) group action is then nothing but a (covariant) functor from G to the category of sets, and a group representation is a functor from G to the category of vector spaces. A morphism between G-sets is then a natural transformation between the group action functors. In analogy, an action of a groupoid is a functor from the groupoid to the category of sets or to some other category.

In addition to continuous actions of topological groups on topological spaces, one also often considers smooth actions of Lie groups on smooth manifolds, regular actions of algebraic groups on algebraic varieties, and actions of group schemes on schemes. All of these are examples of group objects acting on objects of their respective category.

Group Action

Definition Let $(G, \bullet)$ be a group with identity e. Then G is said to act on a set X, if there exists an operator $\star : G \times X \to X$ satisfying the following conditions:

1. $e \star x = x$, for all $x \in X$, and

2. $g \star (h \star x) = (g \bullet h) \star x$, for all $x \in X$ and g, h $\in$ G.

Remark: 1. Let us assume that X consists of a set of points and let us suppose that the group G acts on X by moving the points. Then, the Definition can be interpreted as follows:

(a) the first condition implies that the identity element of the group does not move any element of X. That is, the points in X remain fixed when they are acted upon by the identity element of G.

(b) the second condition implies that if a point, say $x_0 \in X$, is first acted upon by an element h $\in$ G and then by an element g $\in$ G then the final position of x_0 is same as the position it would have reached if it was acted exactly once by the element g $\bullet$ h $\in$ G.

2. Fix an element g $\in$ G. Then the set $\{g \star x : x \in X\} = X$.

For, otherwise, there exist $x, y \in X$ such that $g \star x = g \star y$. Then, by definition,

$$x = e \star x = \left(g^{-1} \cdot g\right) \star x = g^{-1} \star (g \star x) = g^{-1} \star (g \star y) = \left(g^{-1} \cdot g\right) \star y = e \star y = y.$$

That is, g just permutes the elements of X. Or equivalently, each g $\in$ G gives rise to a one-one, onto function from X into itself.

3. There may exist g, h $\in$ G, with $g \neq h$ such that $g \star x = h \star x$, for all $x \in X$,.

Before proceeding further with definitions and results related with group action, let us look at a few examples.

Example: 1. Consider the dihedral group $D_6 = \{e, r, \dots, r^5, f, rf, \dots, r^5f\}$, with $r^6 = e = f^2$ and $rf = fr^5$. Here, f stands for the vertical flip and r stands for counter clockwise rotation by an angle of $\frac{\pi}{3}$. Then D_6 acts on the labeled edges/vertices of a regular hexagon by permuting the labeling of the edges/vertices.

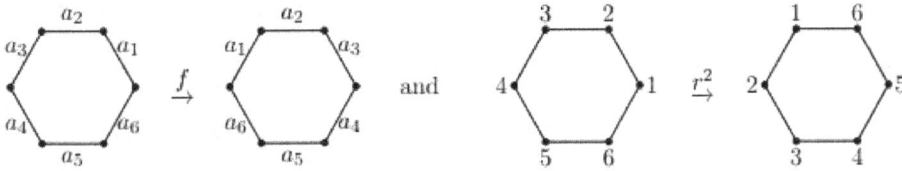

(e) Action of f on labeled edges and of r² on labeled vertices of a regular hexagon.

2. Let X denote the set of ways of coloring the vertices of a square with two colors, say, Red and Blue. Then X equals the set of all functions $h : \{1, 2, 3, 4\} \dashrightarrow \{$Red, Blue$\}$, where the vertices south-west, south-east, north-east and north-west are respectively, labeled as 1, 2, 3 and 4. Then, using Lemma $|X| = 16$. The distinct colorings have been depicted in figure f, where R stands for the vertex colored "Red" and B stands for the vertex colored "Blue". For example, the figure labeled x_9 in figure f corresponds to $h(1) = R = h(4)$ and $h(2) = B = h(3)$. Now, let us denote the permutation (1234) by r and the permutation (12)(34) by f. Then the dihedral group D4 = $\{e, r, r^2, r^3, f, rf, r^2f, r^3f\}$ acts on the set X. For example,

(a) x_1 and x_{16} are mapped to itself under the action of every element of D_4. That is,

$g \star x_1 = x_1$ and $g \star x_{16} = x_{16}$, for all $g \in G$.

(b) $r \star x_2 = x_5$ and $f \star x_2 = x_3$.

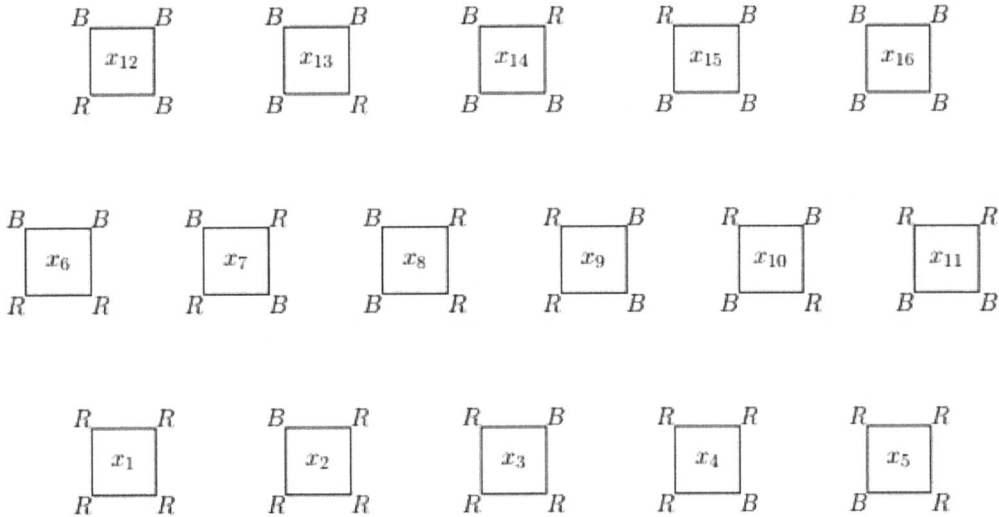

(f) Coloring the vertices of a square.

There are three important sets associated with a group action. We first define them and then try to understand them using an example.

Definition Let G act on a set X. Then

1. for each fixed $x \in X, \mathcal{O}(x) = \{g \star x : g \in G\} \subset X$ is called the Orbit of x.

2. for each fixed $x \in X, G_x = \{g \in G : g \star x = x\} \subset G$ is called the Stabilizer of x in G.

3. for each fixed $g \in G, F_g = \{x \in X : g \cdot x = x\} \subset X$ is called the Fix of g.

Let us now understand the above definitions using the following example.

Consider the set X given. Then using the depiction of the set X in figure f, we have

$$\mathcal{O}(x_2) = \{x_2, x_3, x_4, x_5\}, G_{x_2} = \{e, rf\}, and\ F_{rf} = \{x_1, x_2, x_4, x_7, x_{10}, x_{13}, x_{15}, x_{16}\}.$$

The readers should compute the different sets by taking other examples to understand the above defined sets.

We now state a few results associated with the above definitions. The proofs are omitted as they can be easily verified.

Proposition: Let G act on a set X.

1. Then for each fixed $x \in X$, the set G_x is a subgroup of G.

2. Define a relation, denoted ~, on the set X, by $x \sim y$ if there exists g ∈ G, such that $g \star x = y$. Then prove that ~ defines an equivalence relation on the set X. Furthermore, the equivalence class containing $x \in X$ equals $\mathcal{O}(x) = \{g \star x : g \in G\} \subset X$.

3. Fix $x \in X$ and let $t \in \mathcal{O}(x)$. Then $\mathcal{O}(x) = \mathcal{O}(t)$. Moreover, if $g \star x = t$ then $G_x = g^{-1}Gt\,g$.

Let G act on a set X. Then Proposition 3.6.6 helps us to relate the distinct orbits of X under the action of G with the cosets of G. This is stated and proved as the next result.

Theorem: Let a group G act on a set X. Then for each fixed $x \in X$, there is a one-to- one correspondence between the elements of $\mathcal{O}(x)$ and the set of all left cosets of G_x in G. In particular,

$$|\mathcal{O}(x)| = [G : G_x] = the\ number\ of\ left\ cosets\ of\ G_x\ in\ G$$.

Moreover, if G is a finite group then $|G| = |\mathcal{O}(x)| \cdot |G_x|$, for all $x \in X$.

Proof. Let S be the set of distinct left cosets of G_x in G. Then $S = \{gG_x : g \in G\}$ and $|S| = [G : G_x]$. Consider the map $\tau : S \rightarrow \mathcal{O}(x)$ by $\tau(gG_x) = g \star x$. Let us first check that this map is well-defined.

So, suppose that the left cosets gG_x and hG_x are equal. That is, $gG_x = hG_x$. Then, using Theorem on page number 200 and the definition of group action, one obtains the following sequence of assertions:

$$gG_x = hG_x \Leftrightarrow h^{-1}g \in G_x \Leftrightarrow (h^{-1}g) \star x = x \Leftrightarrow h^{-1} \star (g \star x) = x \Leftrightarrow g \star x = h \star x.$$

Thus, by definition of the map τ, one has $gG_x = hG_x \Leftrightarrow \tau(gG_x) = \tau(hG_x)$. Hence, τ is not only well-defined but also one-one.

To show τ is onto, note that for each $y \in \mathcal{O}(x)$, there exists an h ∈ G, such that $h \star x = y$. Also, for this choice of h ∈ G, the coset $hG_x \in S$. Therefore, for this choice of $h \in G$, $\tau(hG_x) = h \star x = y$ holds. Hence, τ is onto.

Therefore, we have shown that τ gives a one-to-one correspondence between $\mathcal{O}(x)$ and the set S. This completes the proof of the first part. The other part follows by observing that by definition ;

$$[G:G_x] = \frac{|G|}{|G_x|}, \text{ for each subgroup } G_x \text{ of G whenever } |G| \text{ is finite.}$$

The following lemmas are immediate consequences of the above Proposition and Theorem.

We give the proof for the sake of completeness.

Lemma: Let G be a finite group acting on a set X. Then, for each $y \in X$,

$$\sum_{x \in \mathcal{O}(y)} |G_x| = |G|$$

Proof. Recall that, for each $x \in \mathcal{O}(y), |\mathcal{O}(x)| = |\mathcal{O}(y)|$. Hence, using Theorem, one has $|G| = |G_x| \cdot |\mathcal{O}(x)|$, for all $x \in X$. Therefore,

$$\sum_{x \in \mathcal{O}(y)} |G_x| = \sum_{x \in \mathcal{O}(y)} \frac{|G|}{|\mathcal{O}(x)|} = \sum_{x \in \mathcal{O}(y)} \frac{|G|}{|\mathcal{O}(y)|} = \frac{|G|}{|\mathcal{O}(y)|} \sum_{x \in \mathcal{O}(y)} 1 = \frac{|G|}{|\mathcal{O}(y)|} |\mathcal{O}(y)| = |G|.$$

Theorem: Let G be a finite group acting on a set X. Let N denote the number of distinct orbits of X under the action of G. Then

$$N = \frac{1}{|G|} \sum_{x \in X} |G_x|.$$

Proof. By Lemma, note that $\sum_{x \in \mathcal{O}(y)} |G_x| = |G|$, for all $y \in X$. Let $x_1, x_2, \ldots, x_N$ be the representative of the distinct orbits of X under the action of G. Then

$$\frac{1}{|G|} \sum_{x \in X} |G_x| = \frac{1}{|G|} \sum_{i=1}^{N} \sum_{y \in \mathcal{O}(x_i)} |G_{x_i}| = \frac{1}{|G|} \sum_{i=1}^{N} |G| = \frac{1}{|G|} N \cdot |G| = N.$$

Example: Let us come back to Example. Check that the number of distinct colorings are

$$\frac{1}{|G|} \sum_{i=1}^{16} |G_{x_i}| = \frac{1}{8}(8+2+2+2+2+2+4+2+2+4+2+2+2+2+2+8) = 6.$$

Observation: As the above example illustrates, we are able to find the number of distinct configurations using this method. But it is important to observe that this method requires us to list all elements of X. That is, if we need to list all the elements of X then we can already pick the ones that are distinct. So, the question arises what is the need of the above Theorem. Also, if we color the vertices of the square with 3 colors, then $|X| = 3^4 = 81$, whereas the number of elements of D_4 (the group that acts as the group of symmetries of a square) remains 8. So, one feels that the calculation may become easy if one has to look at the elements of the group D_4 as one just needs to look at 8 elements of D_4. So, the question arises, can we get a formula that relates the number of distinct orbits with the elements of the group, in place of the elements of the set X? This query has an affirmative answer and is given as our next result.

Lemma (Cauchy-Frobenius-Burnside's Lemma) Let G be a finite group acting on a set

X. Let N denote the number of distinct orbits of X under the action of G. Then

$$N = \frac{1}{|G|} \sum_{g \in G} |F_g|$$

Proof. Consider the set $S = \{(g,x) \in G \times X : g \star x = x\}$. We calculate $|S|$ by two methods. As the first method, let us fix $x \in X$. Then, for each fixed $x \in X$, G_x gives the collection of elements of G that satisfy $g \star x = x$. So, $|S| = \sum_{x \in X} |G_x|$.

As the second method, let us fix $g \in G$. Then, for each fixed $g \in G$, F_g gives the collection of elements of X that satisfy $g \star x = x$. So, $|S| = \sum_{g \in G} |F_g|$. Thus, using two separate methods, one has),

$\sum_{x \in X} |G_x| = |S| = \sum_{g \in G} |F_g|$. Hence, using the above Theorem, we have

$$N = \frac{1}{|G|} \sum_{g \in G} |F_g|$$

Example: Let us come back to Example. Check that $|F_e| = 16$, $|F_r| = 2$, $|F_{r^2}| = 4$, $|F_{r^3}| = 2$, $|F_f| = 4$, $|F_{rf}| = 8$, $|F_{r^2 f}| = 4$ and $|F_{r^3 f}| = 8$. Hence, the number of distinct configurations are

$$\frac{1}{|G|} \sum_{g \in G} |F_g| = \frac{1}{8}(16 + 2 + 4 + 2 + 4 + 8 + 4 + 8) = 6.$$

It seems that we may still need to know all the elements of X to compute the above terms. To compute $|F_g|$, for any $g \in G$, we just need to know the decomposition of g as product of disjoint cycles.

References

- Lang, Serge (2002), Algebra, Graduate Texts in Mathematics, 211 (Revised third ed.), New York: Springer-Verlag, ISBN 978-0-387-95385-4, MR 1878556, Zbl 0984.00001

- Roth, Richard R. (2001), "A History of Lagrange's Theorem on Groups", Mathematics Magazine, 74 (2): 99–108, JSTOR 2690624, doi:10.2307/2690624

- Rotman, Joseph (1995). An Introduction to the Theory of Groups. Graduate Texts in Mathematics 148 ((4th ed.) ed.). Springer-Verlag. ISBN 0-387-94285-8

- Besche, Hans Ulrich; Eick, Bettina; O'Brien, E. A. (2001), "The groups of order at most 2000", Electronic Research Announcements of the American Mathematical Society, 7: 1–4, MR 1826989, doi:10.1090/S1079-6762-01-00087-7

- Smith, Jonathan D.H. (2008). Introduction to abstract algebra. Textbooks in mathematics. CRC Press. ISBN 978-1-4200-6371-4

- Jahn, H.; Teller, E. (1937), "Stability of Polyatomic Molecules in Degenerate Electronic States. I. Orbital Degeneracy", Proceedings of the Royal Society A, 161 (905): 220–235, Bibcode:1937RSPSA.161..220J, doi:10.1098/rspa.1937.0142

Permissions

All chapters in this book are published with permission under the Creative Commons Attribution Share Alike License or equivalent. Every chapter published in this book has been scrutinized by our experts. Their significance has been extensively debated. The topics covered herein carry significant information for a comprehensive understanding. They may even be implemented as practical applications or may be referred to as a beginning point for further studies.

We would like to thank the editorial team for lending their expertise to make the book truly unique. They have played a crucial role in the development of this book. Without their invaluable contributions this book wouldn't have been possible. They have made vital efforts to compile up to date information on the varied aspects of this subject to make this book a valuable addition to the collection of many professionals and students.

This book was conceptualized with the vision of imparting up-to-date and integrated information in this field. To ensure the same, a matchless editorial board was set up. Every individual on the board went through rigorous rounds of assessment to prove their worth. After which they invested a large part of their time researching and compiling the most relevant data for our readers.

The editorial board has been involved in producing this book since its inception. They have spent rigorous hours researching and exploring the diverse topics which have resulted in the successful publishing of this book. They have passed on their knowledge of decades through this book. To expedite this challenging task, the publisher supported the team at every step. A small team of assistant editors was also appointed to further simplify the editing procedure and attain best results for the readers.

Apart from the editorial board, the designing team has also invested a significant amount of their time in understanding the subject and creating the most relevant covers. They scrutinized every image to scout for the most suitable representation of the subject and create an appropriate cover for the book.

The publishing team has been an ardent support to the editorial, designing and production team. Their endless efforts to recruit the best for this project, has resulted in the accomplishment of this book. They are a veteran in the field of academics and their pool of knowledge is as vast as their experience in printing. Their expertise and guidance has proved useful at every step. Their uncompromising quality standards have made this book an exceptional effort. Their encouragement from time to time has been an inspiration for everyone.

The publisher and the editorial board hope that this book will prove to be a valuable piece of knowledge for students, practitioners and scholars across the globe.

Index